THE INTERNET ON EARTH

THE INTERNET ON EARTH
A GEOGRAPHY OF INFORMATION

AHARON KELLERMAN
University of Haifa, Israel

JOHN WILEY & SONS, LTD

Other Wiley Editorial Offices

John Wiley & Sons Inc., 111 River Street, Hoboken, NJ 07030, USA

Jossey-Bass, 989 Market Street, San Francisco, CA 94103-1741, USA

Wiley-VCH Verlag GmbH, Boschstr. 12, D-69469 Weinheim, Germany

John Wiley & Sons Australia Ltd, 33 Park Road, Milton, Queensland 4064, Australia

John Wiley & Sons (Asia) Pte Ltd, 2 Clementi Loop #02-01, Jin Xing Distripark, Singapore 129809

John Wiley & Sons Canada Ltd, 22 Worcester Road, Etobicoke, Ontario, Canada M9W 1L1

Library of Congress Cataloging-in-Publication Data

Kellerman, Aharon.
    The Internet on earth : a geography of information / Aharon Kellerman.
        p. cm.
    Includes bibliographical references and index.
    ISBN 0-470-84450-7 (alk. paper)
      1. Internet.   2. Web sites – Location.   3. Cyberspace.   4. Information
        services – Location.   5. Information society.   I. Title.

    TK5105.875.I57 K4423 2002
    303.48'33 – dc21
                                                            2002027249

British Library Cataloguing in Publication Data

A catalogue record for this book is available from the British Library

ISBN 0-470-84450-7

Typeset in 10/12pt Lucida Sans by Laserwords Private Limited, Chennai, India
Printed and bound in Great Britain by Antony Rowe Ltd, Chippenham, Wiltshire
This book is printed on acid-free paper responsibly manufactured from sustainable forestry
in which at least two trees are planted for each one used for paper production.

Dedicated to
Yif'at and Hallel, first in a new generation

CONTENTS

PREFACE   xi

ACKNOWLEDGMENTS   xv

1 CONTEXTS   1
  1.1   Information and Knowledge   1
     1.1.1   The information sequence   2
     1.1.2   Information classification   4
     1.1.3   Knowledge classification   6
     1.1.4   Information, knowledge, and geography   7
  1.2   The Information Society   9
     1.2.1   Processes leading to the information society   11
     1.2.2   Information-rich society (1960s–1970s)   12
     1.2.3   Information-based society (1980s–1990s)   13
     1.2.4   Information-dominated society (1990s–2000s)   14
  1.3   The Information Economy   15
     1.3.1   Major elements of the information economy   16
     1.3.2   Media   18
     1.3.3   Operators   19
     1.3.4   Information economic geography   21
  1.4   Information Politics   22
  1.5   Information Law   25
  1.6   Conclusion   26

2 BASICS   29
  2.1   The Scope of Information Geography   29
  2.2   Space   31
     2.2.1   Social space   31
     2.2.2   The real and the virtual   33
  2.3   Place   39
     2.3.1   Interpretations of place   39
     2.3.2   Globalization   41

CONTENTS

      2.3.3 The global and the local   43
      2.3.4 Place production   49
  2.4 Conclusion   50

3 TECHNOLOGY   **53**
  3.1 Information and Technology   53
  3.2 Technology and Flows   54
  3.3 Knowledge ⇒ Innovation ⇒ Technology   57
      3.3.1 Knowledge for innovation   58
      3.3.2 Regions of knowledge and innovation   60
      3.3.3 The innovation process   62
      3.3.4 The innovation process and the regional economy   64
  3.4 Information Technology Regions   68
      3.4.1 IT regions worldwide   68
      3.4.2 IT R&D and production in selected countries   74
  3.5 Conclusion   79

4 PRODUCTION   **81**
  4.1 Information Volumes and Origins   81
      4.1.1 How much?   81
      4.1.2 From where?   83
  4.2 The Internet: Evolution and Structure   85
      4.2.1 Essence   85
      4.2.2 Evolution   87
      4.2.3 Structure   89
  4.3 A Conceptual Framework for Information Production   91
      4.3.1 Process   91
      4.3.2 Location   92
  4.4 Ranking Urban Centers of Information Production   95
      4.4.1 A global view   96
      4.4.2 The U.S.   97
      4.4.3 Europe   98
      4.4.4 Asia   100
  4.5 Global Centers: New York and Los Angeles   101
      4.5.1 Global centrality   101
      4.5.2 New York   103
      4.5.3 Los Angeles   106
  4.6 IT R&D and Information Production   109
  4.7 Conclusion   110

5 CONTENTS   **113**
5.1   Content Demand and Location   113
5.2   Capital as Information   119
   5.2.1   Capital, space and information   119
   5.2.2   Flows   120
   5.2.3   Management   122
   5.2.4   Urban expressions   124
5.3   E-Commerce and Location   127
   5.3.1   Nature   127
   5.3.2   Structure   128
   5.3.3   Geography   131
   5.3.4   Vendors   132
5.4   Geographic Language   133
   5.4.1   Space, place and communications media   133
   5.4.2   Simulation   135
   5.4.3   Language and metaphor   136
   5.4.4   Virtual communities   137
5.5   Conclusion   138

6 TRANSMISSION   **139**
6.1   The Internet Backbones   140
   6.1.1   Definitions   140
   6.1.2   The U.S.   141
   6.1.3   Europe   146
   6.1.4   Global view   146
   6.1.5   Switching facilities   148
   6.1.6   Hosting facilities   152
6.2   Flows   153
6.3   U.S. Leadership in Telecommunications   155
   6.3.1   The historical development of U.S. leadership   155
6.4   Conclusion   158

7 MEDIA   **161**
7.1   Leading Nations   161
   7.1.1   Relative advantage in telecommunications
   *   at the national level*   162
   7.1.2   Leading nations in 1975   163
   7.1.3   Leading nations in 1986   165
   7.1.4   Leading nations in 1995   166

CONTENTS

		7.1.5	Leading nations in 1999	167
		7.1.6	Factors of leadership	171
	7.2	The Digital Divide		176
		7.2.1	Definition	176
		7.2.2	International	177
		7.2.3	Intranational	180
	7.3	Conclusion		181

8 | **CONSUMPTION** | | | **185**
	8.1	Social Uses of the Internet		186
		8.1.1	Networking	186
		8.1.2	Individualism	188
	8.2	Internet Consumption in Cities		188
		8.2.1	Penetration rates in American cities	188
		8.2.2	Penetration rates in Asian cities	191
		8.2.3	Use	192
	8.3	Use and Location		193
		8.3.1	Geolocation	194
		8.3.2	Wireless information transmission	195
	8.4	Broadband		198
		8.4.1	Diffusion by country	199
		8.4.2	Penetration rates by cities	201
	8.5	Conclusion		102

9 | **BEYOND** | | | **205**
	9.1	Challenges		205
	9.2	Geography of Information		208
	9.3	Geography of Information?		215

| **REFERENCES** | | | | **217** |

| **GLOSSARY OF ABBREVIATIONS** | | | | **239** |

| **INDEX** | | | | **241** |

PREFACE

'If information is power, whoever rules the world's telecommunications system commands the world'

(Hugill 1999, p. 2)

The study of the spatial dimensions of information is rather new, emerging as of the early1990s. The growing interest in this area may be traced in the information explosion accompanying the introduction of the Internet as a globally accessible and comprehensive information system. The very introduction, as well as the rapid growth and diffusion of the Internet, as of the mid-1990s, became feasible through major technological innovations and transitions in the two preceding decades. These technological innovations and transitions were concentrated in *information and communications technology* (ICT), or in the electronic processing and transmission of information. It would have been impossible for information to be so heavily produced, transmitted and consumed, as we currently experience, if not for the previous and continuous extensive developments in ICT.

The geographical study of information and telecommunications followed this same process, so that, beginning in the early 1980s, the spatial study of telecommunications emerged, followed later on by the study of information. The direction taken by many geographers for the study of telecommunications has been criticized as treating communications similar to 'pipelines made of concrete or steel' (Hillis 1998, p. 553), not accentuating enough either the social and linguistic dimensions of communications (as the vehicle), or the substance moving through the 'pipelines' (information). Interestingly enough, the geographical study of information per se, notably of cyberspace, has picked up exactly from this criticism, namely it has focused on the social and cultural dimensions of information. This book attempts to add to this accumulating knowledge a systematic study of basic geographical aspects of information, focusing on the Internet: technology, production, contents, transmission, media, and consumption.

The major objective of this book is to develop and present a systematic geography of information within the context of information society and the Internet, or in other words, to put forward a 'conventional' geography on a slightly unconventional geographical topic, information. Linked to this major objective is the challenge to address and discuss the notion of information as an object for geographical study, with its own spatial dimensions. The book will discuss general concepts regarding the geography of information, and will outline a specific geography of information, that of the Internet, which represents electronic information at large, and which constitutes the most powerful and comprehensive information system currently in global use. A third objective is to highlight some of the interrelationships between information and other systems, notably cities as spatial entities, and finances and commerce as economic activities.

My own interest in the geography of information has risen out of my interest in the geography of telecommunications, to which I devoted my volume *Telecommunications and Geography* (1993). This interest has benefitted tremendously from the emerging community of information and telecommunications geographers, and their annual meetings, organized through both the *E**Space* project, initiated and headed by Kenneth E. Corey and Mark I. Wilson, as well as through the *International Geographical Union* (IGU) commissions, first the one on telecommunications, headed by Henry Bakis, followed in 2000 by that on the information society (headed by myself).

I received financial support from the National Science Foundation, through the research project on cities and telecommunications, carried out at the *Taub Urban Research Center New York University*, and led by Mitchell Moss and Anthony Townsend. It sponsored my studies on the production of Internet information, and on the global becoming local. Developing this book project was directly supported by the *Burda Center for Innovative Communication*, at the Communications Department, *Ben-Gurion University*. My fellow information geographers from literally all over the globe have assisted me in most generous and dedicated ways, through the provision of data and reports, as well as through the sharing of research findings. These colleagues and friends are too numerous to be all mentioned here; I owe them all much gratitude. Two of my colleagues, Kenneth E. Corey and Barney Warf read several draft chapters of the book and provided me with their valuable comments and criticisms. The illustrations were produced by Miri Yehuda and Noga Yoselevich, both at the *University of Haifa*, and I am grateful to them for their patience and dedication. All those who shared with me information, knowledge and expertise, do not share with me, however, in the responsibility for the contents of this volume.

Being involved in heavy administrative duties during the writing period of this book, I owe much to my wife, Michal, and to my children Tovy, Eli, Miri, and Noga for bearing with my long hours of research at unusual times. This book is dedicated to the new generation of my family, headed by my granddaughters, Yif'at and Hallel.

ACKNOWLEDGMENTS

The following sections have been previously published in my papers with some changes: 'Information classification'; part of 'Location'; and 'Content demand and location' in Where does it happen? The location of the production and consumption of Web information. *Journal of Urban Technology* **7**, 2000: 45–61; 'Information society' in Phases in the rise of information society. *Info* **2**, 2000: 537–541; 'Media', 'Operators', and parts of 'U.S. leadership' in Fusions of information types, media, and operators, and continued American leadership in telecommunications. *Telecommunications Policy* **21**, 1997: 553–564; 'The innovation process and the regional economy'; Table 3.6, in Conditions for the development of high-tech industry: The case of Israel. *Tijdschrift voor Economische en Sociale Geografie* **93**, 2002: 270–286;. 'Global centers: New York and Los Angeles' in The global leadership of New York and Los Angeles in information production. *Journal of Information Technology* **9**, 2002: 21–35; 'The use of geographic language' in Space and place in Internet information flows. NETCOM **13**, 1999: 25–35; 'Leading Nations' in Leading nations in the adoption of communications media, 1975–1995. *Urban Geography* **20**, 1999: 377–389.

CONTEXTS

1

'Even in the information economy, geography matters!'

(Goddard 1990, p. xvii)

The Internet on Earth: A Geography of Information, the title of this book, sounds, hopefully, straightforward. Read it again now: *The Internet, on earth! A geography of information?* The meaning of the title now looks almost the opposite of the first reading. The following chapters attempt to convince those who tend to read the title in its second punctuation, to accept the first one. Thus, we shall go through various of their components in order to develop a geography of the Internet/information.

The geography of information is interrelated with other aspects and dimensions of information. This chapter will present some of these, beginning with differentiations among several types of information, and following with elaborations on information society, economy, politics, and law. These discussions will highlight, among other things, related spatial aspects.

1.1 Information and Knowledge

Information is a term used in ambiguous ways, notably since the introduction of information technology, which has permitted the storage, processing and transmission of enormous quantities of information in electronic formats. On the one hand, it refers to a wide family of communicative, mostly codified, materials, which include data, information as a class, knowledge and innovations. As Roszak (1991, p. 13) noted: 'in its new technical sense, *information* has come to denote whatever can be coded for transmission through a channel that connects a source with a receiver, regardless of semantic content'. On the other hand, the term information is also used for a specific class of communicative materials, to be defined in the next section. Information at large is obviously something intangible, although its containers or media have traditionally been material objects, mostly in the form of paper products, such as books, magazines, letters, documents, lists, etc. The emergence of electronic transmission and storage media, such as radio, TV, cassette recorders, followed later on by computers and the Internet, has once again accentuated the intangible and abstract character of information.

1.1.1 The information sequence

The communicative materials, data, information, knowledge and innovation, have each received numerous independent definitions. However, it is also possible to refer to them as a sequence, in which data lead to the production of information, which, in its turn, may lead to the development of knowledge, and vice versa. Knowledge, for its part, may bring about the development of innovations (see Malecki 2000a, p. 104). We shall first review some of the definitions for each of the communicative materials.

Data: 'A series of observations, measurements, or facts in the form of numbers, words, sounds and/or images. Data have no meaning but provide the raw material from which information is produced' (Roberts 2000).

Information: Information per se, as a class of communicative materials, has received numerous definitions, estimated at over 100, proposed in about 40 disciplines (see e.g. Machlup 1983, Braman 1989). It was claimed *to be* an activity, a life form and a relationship (Barlow 1994). It was further seen *as* resource, commodity, perception of pattern, and a constitutive force in society (Braman 1989). From a sequential perspective, information is 'data that have been arranged into a meaningful pattern. Information must relate to a context for it to have meaning' (Roberts 2000, see also Borgmann 1999). Porat (1977, p. 2) similarly defined information as 'data that have been organized and communicated' (see also Castells 2000, p. 17).

Knowledge: Definitions of knowledge refer to its relation to information. It has, thus, been defined as 'the application and productive use of information. Knowledge is more than information, since it involves an awareness or understanding gained through experience, familiarity or learning' (Roberts 2000). For Roszak (1991, p. 105) 'ideas create information, not the other way around. Every fact grows from an idea'. Other commentators, however, such as Boisot (1998, p. 12), argue that 'knowledge builds on information that is extracted from data'. A kind of two-way relationship between knowledge

2

and information was illuminated by Neil Postman (1999, p. 93) as follows: 'I define knowledge as organized information, information that has a purpose, that leads one to seek further information in order to understand something about the world. Without organized information, we may know something *of* the world, but very little *about* it. When one has knowledge, one knows how to make sense of information, knows how to relate information to one's life, and, especially knows when information is irrelevant'.

Bell (1976, p. 175; see also Castells 2000, p. 17) added to knowledge the communications element: 'knowledge is a set of organized statements of facts or ideas, presenting a reasoned judgment or an experimental result, which is transmitted to others through some communication medium in some systematic form'. On the other hand, knowledge may also be created without information and communications: 'Information is acquired by being told, whereas knowledge can be acquired by thinking... Thus, *new knowledge can be acquired without new information being received*' (Machlup 1983, p. 644). By the same token, 'information transfer is always necessary to knowledge exchange, the reverse is not always true' (Storper 2000b, p. 56).

There are communicative materials which are difficult to classify as either information or knowledge, such as the so-called *meme*. An example in this regard are work practices, norms and conventions transferred through behavior and imitation (Storper 2000b, pp. 56–57). Similarly, Thrift (1985) discussed knowledge availability to humans, through both individual and societal social action, defining knowledge simply as 'information about the world' (p. 367).

Innovation: It may be defined as the creation of new knowledge, through an intrinsically uncertain problem-solving process based on existing knowledge and/or information. Innovative knowledge may lead to the introduction of innovative products or the application of a novel production process, either through radical breakthroughs or through incremental improvements (Feldman 1994, p. 2, Feldman 2000, pp. 373–375). Knowledge may thus be viewed as an asset, which serves as an input (competence) leading to an output, in the form of innovation, which may thus be viewed as another kind of knowledge (OECD 2000a, p. 13). The role of innovation has become of major contemporary importance, since about 80% of productivity growth in advanced countries has been attributed to innovation (Sternberg and Arndt 2001).

Following the definitions for the four communicative materials, we may recognize four basic sequential processes for information at large, accentuating its transformative and communicative nature (Figure 1.1). In the first,

3

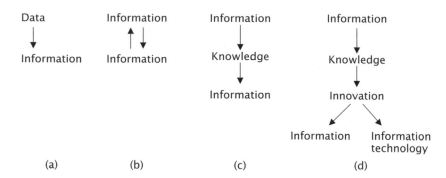

Figure 1.1 The information sequence

data are turned into information through meaningful patterns and context. In the second, information yields information, such as in a spoken or written exchange between two or more people. In the third, information turns into knowledge, through its application and use. However, knowledge may yield information, and knowledge is required for the additional development of information (Roberts 2000). Innovation, or the creation of new knowledge, notably of applied nature, is dependent on previously existing knowledge, tacit as well as codified, and information. On its part it may yield new information technology, as well as new information.

1.1.2 Information classification

There are various ways to classify information as a class of communicative materials. One simple classification is by contents: social, business, entertainment, news, educational. Traditionally, there have evolved close ties among several of these information types, through the technologies used for their production or transmission. Thus, books and articles share similar printing technologies, entertainment and news are transmitted through radio and television, movies also include voice and music, and the telephone is used for both business and social conversations.

Computer-based multimedia technology was noted by Castells (1996, pp. 371–372) as enabling an 'integration of all messages in a common cognitive pattern'. While the messages of an educational movie may differ from that of a news broadcast, their operational contexts may be similar, using icons, video-games, audio-visual shows, etc. Thus, multimedia can be used to present a wide variety of cultural expressions. The most integrative information technology is the Internet. This computerized information system functions as a comprehensive information store, containing all types of information, regardless of contents or form.

Another most basic classification is by form: text, data, graphics, voice, and picture. From the perspective of information production and transmission, major developments in transistor and computer technologies as of the 1970s, have turned all these information types into similar digital bit formats, differentiated mainly by the channel bandwidth required for their transmission. From the perspective of information users this classification has lost much of its significance through the introduction of the Internet in the 1990s, permitting integrative transmissions of information in every form through a single channel, transforming all information forms into electronic signals (see Kellerman 1997).

A third simple classification of information has focused on information transmission, distinguishing between information producers or senders, transmission media or telecommunications channels, and information receivers or users. However, as far as the Web system of the Internet is concerned, information may be produced and installed on the system, but not necessarily consumed by anybody. In other words, if nobody calls for a certain site, it will neither be transmitted nor received or used. Thus, electronic information has to be classified into designated and undesignated. Designated information is information which is transmitted to specific individual receivers (such as telephone calls or fax transmissions) or to specific audiences (e.g. CATV), whereas undesignated information is put on a transmission device (i.e. a host), and potentially received by any number of users.

An alternative classification which would fit mostly electronically transmitted information may look at it as something that does not stand for itself, but is rather woven into the wider context of social or economic processes. Information constitutes a resource or product, which may thus either be free of charge or involve a price for its use or purchase (see also Zook 1998). Four types of information may be recognized by economic process (Kellerman 2000a):

(1) Pure information: Personal, academic, and some business information, in the form of e-mail, or website information, is provided at no charge (other than for its transmission), and with no intention to create or promote purchases. Thus, some of this information is designated (e-mail), and some is not (websites).

5

(2) Information as part of sale processes: Information transmission over the Internet has often become *part* of the sale process of both products (e.g. books) or services (e.g. airline tickets). This type of information is undesignated in most cases. However, it may turn designated, once a purchase is made (e.g. order confirmation or progress reports). Such types of information are currently part of the marketing systems of almost all material products, as well as of many services.

(3) Information as product: Information itself may be sold over the Internet (e.g. astrological forecasts) to any undesignated customers. Here too, once a sale has been executed, the information transmission becomes designated.

(4) Products transformed into information: Probably the only real product transformed into electronic information is also the most powerful one – money. Capital as a resource or as a product is currently transmitted almost exclusively over telecommunications channels, not necessarily over the Internet. The transmission of capital is obviously highly and most clearly designated.

The transmission of pure information through voice telephony, as of 1876, and later on of text and graphical transmission through fax (type 1), paved the way for the transmission of capital in the form of data as of the 1970s (type 4). The construction of dense domestic and international telecommunications networks facilitated the evolution of the Internet as of the late 1980s–early 1990s, so that electronic information could be integrated into sale processes (type 2), and information itself could be sold (type 3). The emergence of the Internet has enhanced the transmission of pure information and capital, making possible their transmission in more sophisticated ways than before.

1.1.3 Knowledge classification

There are numerous classifications of knowledge (for a review, see Howells 2000). The classification of knowledge dates back to Aristotle, who distinguished between three types of knowledge: *èpistemé*, or theoretical and universal; *techné*, instrumental knowledge, practice related, and context specific; and *phronesis*, experience-based, normative, related to common sense and context specific (Johnson and Lundvall 2001, p. 6). A recently proposed classification of knowledge into four types may be partially related to this ancient classification: know-what (facts); know-why (principles, similar to *èpistemé*); know-how (skills, similar to *techné*); and know-who (socialization) (Johnson and Lundvall 2001).

Another dominant classification of knowledge is into codified, or explicit, and tacit, or implicit (Cowan *et al.* 2000, Roberts 2000). Codified knowledge is recorded or transmitted through symbols (letters and letters turned into words and sentences, as well as digits and drawings), whereas tacit knowledge develops through learning, and is, thus, similar to know-how (skills). These two forms of knowledge complement each other, and they are normally present in any specific piece of knowledge. Codified knowledge can be converted into information and is mostly viewed, therefore, as more easily transferable through information technology, whereas tacit knowledge is more readily transferred through the transfer of knowledge-bearers, namely people (see Cowan *et al.* 2000, Roberts 2000, Johnson and Lundvall 2001). It is arguable, though, whether tacit knowledge is becoming less important than codified knowledge in an era of proliferating information technology, notably because 'tacit knowledge remains a prerequisite in *all* knowledge activities' (Howells, 2000, p. 61). Common professional education of people from different countries makes them talk the same professional language, thus assisting in the transfer of tacit knowledge (Storper 2000a, p. 157). As we shall note in Chapter 3, tacit knowledge is a crucial key for the innovation process in the development of new technologies.

1.1.4 Information, knowledge, and geography

From a spatial perspective, information at large might be viewed as a rather *active* action that occupies time but does not occupy space at all (Barlow 1994), or it may be defined as something that is only *passively* contained within one's mind, when viewed as 'anything that is known by somebody' (Machlup 1980, p. 7). On the other hand, information might be seen as something that is constantly moving over space: 'information that isn't moving ceases to exist as anything but potential' (Barlow 1994, p. 7). The spread or movement of information differs from the movement of material objects in that 'it leaves a trail everywhere it's been... information can be transferred without leaving the possession of the original owner' (Barlow 1994, p. 7). Thus, its impact on places may potentially be much more significant, when compared with those of other entities which are fully moving or being fully moved from one place to another.

In terms of location, information has been defined as a 'compromise between presence and absence', since it represents a 'form of something without the thing itself' (Latour 1987, p. 243). Communicating information is, thus, 'being; persons literally occupy the media they use; their existence cannot be separated from these symbolic systems' (Adams 1995).

7

Information, though being abstract and most flexible for electronic process-ing and transmission, has a simultaneous presence in places of origin and destination. Towards its transmission over space, it may be packaged and transformed. Furthermore, its interpretation and uses may change from one receiver to another, so that the same piece of information may embed differ-ently in various places, thus bringing about different aggregate geographical patterns.

These various statements or definitions point to three major geographical aspects of information at large, similar to other human products: production, flow, and consumption or dissemination. None of these aspects is new or related in principle solely to the contemporary introduction of information technology. This technology rather has changed the scale and form of infor-mation production and consumption, as well as the speed of its transmission. If the major traditional container of information was paper, followed later on by films and electro-mechanical signals (through the telephone), more recent information technology has made it possible to transform all informa-tion forms into standard electronic bits, and has offered a singular channel for its transmission, the Internet. In principle, these innovations should have made the production, transmission, and consumption of information at large, ubiquitous.

The production of information, in the narrower sense of the term, differs, however, from place to place. It is not only dependent on the size of a place but also on the types of economic activities taking place and concentrated in it. Thus, the finance industry is a major producer of information, so that cities which specialize in finance tend to produce more information than others (see Chapter 4). Since the development of this industry had its roots long before the introduction of information technology, the initial and cumulative advantages in this industry have created a rather early differentiation in infor-mation production. As far as knowledge production is concerned, notably technology production, the differentiation may be even more significant, with fewer places dominating the map of technology production, with an initial and cumulative advantage obtained by cities with universities specializing in technological studies and research (see Chapter 3).

The transmission of information is dependent, foremost, on the availabil-ity of a transmission system. The more bandwidth it provides, the more and faster transmission of information it may handle. Transmission systems refer, thus, to information in terms of its size, rather than its content, pric-ing transmission by volume or time rather than value by content. The price of information at large may differ sharply by content, notably through the difference between information and knowledge. The transmission of information is often considered a global phenomenon, constrained no longer by political boundaries, though this is not always the case. However, legal procedures, notably those relating to copyrights may regulate the flows of various types of information, such as movies, and much of the flows of knowledge. It is also much easier to move codified rather than tacit knowledge, though conferences and e-mail permit an accelerated transmission of this form of knowledge as well (Roberts 2000).

The consumption and dissemination of information and knowledge also may differ from one place to another, and again, not necessarily because of differences in population size. There is a need for some socio-cultural similarity or *contextuality* between the sender and receiver of information in order for a significant dissemination of information to take place (Johnson and Lundvall 2001). Such contextuality relates to major aspects such as language, education, and economic development. This is notably striking as far as the transmission of knowledge is concerned. The transmission of rather codified knowledge depends on the level of tacit knowledge at the receiving station for its dissemination (Johnson and Lundvall 2001).

The spatial dimension of knowledge, mostly as an economic resource for the recently evolving knowledge-based economy, has been highlighted by numerous authors in both economics and geography, and several readers have been recently published in this subject (Dunning 2000a, Bryson *et al.* 2000, and partially Clark *et al.* 2000). Economic knowledge has become of rather general importance, not just for high-tech industries (Bryson *et al.* 2000, pp. 1–2; see also Hodgson 1999, pp. 181–182). The elaboration of the locational and flow aspects of knowledge and innovation will be treated in Chapter 3, devoted to technology. As a conclusion, it would suffice here to note that information and knowledge are neither concentrated in one place, nor universally distributed. They present rather complex geographies of production, flows and consumption, dependent on and regulated by various factors, notably technology, social and cultural aspects, economic development, and legal procedures.

1.2 The Information Society

The term *information society* has been increasingly in use since the early 1980s, though earlier terms such as the *age of information* date back to the early 1970s (Kellerman 2000b). These and other terms have emerged within the context of numerous attempts to coin societal transformations since the early 1950s. Beniger (1986, pp. 4–5) counted 75 such terms proposed between 1950 and 1984, almost all of which have not been adopted for the naming of our current age.

Definitions for information society highlight two of its major facets, namely the economic and the cultural. At the economic end 'in an information society, information is the most important commodity' (European Commission 1996, p. 7), whereas culturally it is 'a society that brings about a general flourishing state of human intellectual creativity, instead of affluent material consumption' (Masuda 1980, p. 3). Castells (1998, p. 67) related these two facets by claiming that information society 'is based on the historical tension between the material power of abstract information processing and society's search for meaningful cultural identity'. Castells (2000, p. 21, n. 31) also differentiated between *information society* and *informational society*. Whereas the first relates to the role of information in society, which has always been of some importance, the second relates to 'a specific form of social organization in which information generation, processing, and transmission become the fundamental sources of productivity and power because of new technological conditions'.

The information society constitutes two major processes, production and consumption. These do not necessarily have to develop to high levels in all aspects of production and consumption of information in a given national information society. At the production end, several things may be produced. One major production process may be the innovation and wide-scale production of the hardware of information society, such as computers and telecommunications devices and equipment. Another major production process may be computer software, and a third one may be information itself, notably electronic, such as Internet sites, television programs and movies.

High levels of consumption of information may too be expressed in both hardware, software and information per se. The wide adoption of telecommunications and information devices such as PCs, telephones, TVs, etc., is one indicator. Sales of software are another, as are the number and duration of domestic and international phone calls, or the proportion of homes connected to cable TV, or to the Internet.

The rise of information society during the last three decades has sometimes been viewed, explicitly or implicitly, as a single phase process (see e.g. Schement 1989, Webster 1994). Some still consider the term *information society* to be a concept rather than a mature phenomenon, claiming that Western society is still in a process of *informatization*, thus lacking clear

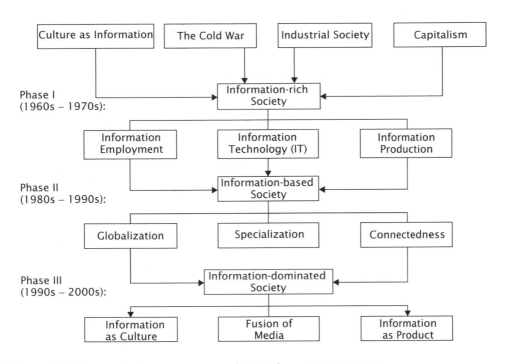

Figure 1.2 Phases in the emergence of the information society

quantitative measures to demonstrate the existence of information society (Halal 1993, Martin 1995). The following discussion outlines three stages in the development of information society: information-rich; information-based; and information-dominated (Figure 1.2). The incubation of the information society took place chiefly in the U.S., within four major societal elements: industrial society; capitalism; the Cold War; and culture as information.

1.2.1 Processes leading to the information society

Some commentators recently have argued that information as a major power has continuously transformed American society, since the very beginnings of European settlement in North America (Chandler and Cortada 2000). The emergence of information society, and as we shall see in Chapter 4, the largest concentration of information production, has been in the U.S. Whether industrial society has given way to information society by continuity or change, has been extensively debated in the literature (see e.g. Gottmann

11

1961, Bell 1976, Kellerman 1985; 1993a; Schement 1989, Masuda 1980, Lyon 1995, Castells 1996, p. 228). On the one hand, the development of information technology (IT) with the consequent need for skilled workers who consume and produce more information, is viewed as nesting within industrial society and the subsequent service economy. On the other hand, information technologies allow 'a direct, on-line linkage between different types of activity in the same process of production, management, and distribution, establish a close, structural connection between spheres of work and employment artificially separated by obsolete statistical categories' (Castells 1996, p. 228). Thus, at a certain stage, information and knowledge may replace labor and possibly also capital as leading production factors.

Capitalism has been recognized as a major force in the emergence of an information society by facilitating the transfer of information from the public to the private sector, or 'the privatization of information' (Schiller 1981, Schement 1989). This process involved the rise of a class of technocrats in the so-called *programmed society*, in conflict with more disparate groups being governed or managed by them (Touraine 1974, Lyon 1995).

A third major force in the rise of information technology and society in the U.S. was the Cold War (Nye and Owens 1996, Lyon 1988, Masuda 1980). Computers were widely required for missiles and defense from missiles, as well as for space exploration. Furthermore, the contemporary Internet network had its roots in the earlier Pentagon ARPANET e-mail system.

A fourth major dimension facilitating the emergence of information society is the nature of cultural activity in developed societies, requiring a constant and rich transmission of written and oral information which used to be constrained by volume, form, time and distance. Thus, the very accent on information and information transmission in contemporary information society was not novel.

1.2.2 Information-rich society (1960s–1970s)

With these four dimensions in place, the first phase in the evolution of information society took place between the 1960s and 1970s, and may be termed as the *information-rich society*. The characteristics of this phase included growing emphases on information production and employment in information work, mainly vis-a-vis the introduction of information technologies.

Growing employment in information-related activities has been recognized as an early and central aspect of information societies: 'The term "information society" has been used to describe socioeconomic systems that exhibit high employment of information-related occupations and wide diffusion of information technologies' (Katz 1988, p. 1). Porat (1977) and Machlup (1962) showed that a large proportion of U.S. workers were *information workers*, employed in information-related occupations, or in the so-called *knowledge sector*, already back in the 1960s, followed in the 1970s to 1980s by other leading economies (Hepworth 1990), and growing significantly since then (Leinbach 2001). However, it was not until later phases in the emergence of information society that information became a common thread in production as well as consumption, through the introduction of personal computing and a common channel of information flow, the Internet.

The introduction of information technology via computers and telecommunications and its rapid and wide diffusion and adoption have become the driving force of information society, interrelated with the growth in information employment. The impact of information technology over the years has been on two levels. On the one hand, it has been a major enabling force for society, allowing inexpensive recording and storage, and fast processing and transmission of information (Martin 1988, p. 40, Steinfeld and Salvaggio 1989, pp. 2–3, European Commission 1996). It has also become an extensive industry in itself, namely the high-tech industry, characterized by intensive R&D and entrepreneurship.

A third major aspect of the information-rich society of the 1960s–1970s was the growth in information production. It seems obvious that the introduction of information technology and the growth in employment in information would lead to the production of larger volumes of information. However, this growth also had to do with the expansion of research and study in universities and research institutes during these decades, yielding ever increasing numbers of books and journals (Bell 1976, Steinfeld and Salvaggio 1989).

1.2.3   Information-based society (1980s–1990s)

The growth in information volume, technology and employment led to a second phase in the emergence of information society, the *information-based society* of the 1980s–1990s. This phase has been characterized by three trends, all based in their part on developments in the first phase: globalization, specialization and connectedness.

The ability to move information instantaneously worldwide has become possible with the rapid development of international telephony, the Internet, and cable/satellite television. These technological breakthroughs have reduced the significance of international boundaries to the movement of

13

information, so possibly weakening statism. This has been evident in various economic, social and cultural areas, including news coverage, banking, commerce, and social contacts. The pace of information production and interaction has quickened, with the shrinking of space (friction) in time.

The second phase in the rise of information society has been typified also by the rapid diffusion and adoption of *information devices*, such as telephones, cellular telephones, fax machines, personal computers, television sets and the like (Kellerman 1999a). Special appliances have evolved for specific uses of information, together with specialized suppliers of equipment, software and information.

A third characteristic of information-based society has been increased connectivity. Internet technologies have permitted the recording and transmission of all forms of information, namely text, data, graphics, voice and pictures in electronic digital bit format. Coupled with fast and low-priced telecommunications and PCs they have allowed an increased connectedness of individual customers with service providers, as well as complex connections among goods producers and service providers among themselves (Kellerman 1997). Furthermore, the interrelationship between electronic information and printed information has changed with the introduction of low-priced high-quality laser printers and optical scanners, permitting non-professional users to produce high-quality paper information products. These developments have rendered society in developed countries increasingly information-dependent (Feather 1994, p. 2, Kitchin 1998, p. 131).

1.2.4 Information-dominated society (1990s–2000s)

Unfolding in the late 1990s towards the 2000s, has been a third phase in the rise of information society which might be termed *information-dominated society*. Information production, transmission and use has become a leading if not *the* leading economic and social activity, both as a product in itself and as a service leading to the production or consumption of material products. As such, three additional characteristics have been added to information society: information becoming a major product, information media beginning to fuse into each other, and information becoming a culture.

Information has increasingly become a commodity in its own right. By the late 1990s, revenues from the sale of information were matching those from the sale of material products and services. Major examples are the sale of data sets relating to Internet users, or the tremendous growth in the sales of software and TV programs. The U.S. has become the world leader in the sale and distribution of electronic information (Kellerman 1997) (see Chapter 6).

14

Liberalization trends in the provision of information services to house-holds, as well as technological advances, have brought about early signs of possible fusions among different forms of information, their transmission and use. Thus it has become possible, for example, to use the computer also as a telephone, fax and TV, and receive several of these services from a single service provider. This fusion may possibly mature into a single appliance for information consumption and production, as well as so-called *public networks* of data and software (Halal 1993, Kellerman 1997).

The information society becomes, at this third phase of its development, a society with a culture of information. Some commentators have seen this as 'recognition of the cultural value of information through the promotion of information values in the interest of national and individual develop-ment' (Martin 1988, p. 40). Others believe that beyond promoting the proper and aesthetic production, transmission and consumption, the culture of information may turn into one of power:

> *Cultural battles are the power battles of the Information Age. They are primarily fought in and by the media, but the media are not the power-holders. Power, as the capacity to impose behavior, lies in the networks of information exchange and symbol manipulation, which relate social actors, institutions, and cultural movements, through icons, spokespersons, and intellectual amplifiers...Culture as the source of power, and power as the source of capital, underlie the new social hierarchy of the Information Age. (Castells 1998, p. 348).*

Another important cultural dimension of information is the changing sig-nificance of time and space. Instantaneous written and oral communications in global space intensify the pace of work and alter working times. In terms of cultural symbolism and reality, our source and anchor have gradually been in a change from traditional national territory to a global virtual one, bringing Castells (1998) to declare that 'the space of flows of the Information Age dominates the space of places of people's cultures' (p. 349). But others see no cultural imperialism in the blurring of national boundaries and no threat to domestic democratic institutions (Kitchin 1998, pp. 101–102, see also Graham 1997).

1.3 The Information Economy

It was geographer Jean Gottmann who first identified the emergence of an urban information economy in the U.S., in his 1961 seminal *Megalopolis*, in

15

which he defined the *quaternary occupations* as 'those supplying services that require research, analysis, judgment, in brief, brainwork and respon-sibility' (p. 580). Abler *et al.* (1977, p. 200) broadened the scope of the quaternary sector into *information activities*. The quaternary sector was more widely defined by Bell (1976), to include trade, finance, insurance and real estate (see also Kellerman 1985). The roots of the contemporary information economy lie, therefore, in the post-industrial economy and the service economy identified in the 1970s–1980s. The concept *information economy* was first coined by Porat (1977, see also Hepworth 1990, p. 6), when he commented on the growing shares of information in U.S. GNP and labor force.

1.3.1   Major elements of the information economy

Information may be considered as both a product and a resource or input for various uses by both individuals and businesses. It constitutes a prod-uct in the most simple way when it is sold, and it serves as a resource or input in many ways, such as information on market trends being used as an input for the production of material products. Goddard (1990, 1992) identi-fied four characteristics of the evolving information economy: the increasing centrality of information in the production of goods and services; the devel-opment of information technology; information becoming a commodity in itself; and economic globalization. The very incorporation of information into the production of goods and services is not novel (Miles and Robins 1992, p. 2), bringing some to comment that 'every business is an informa-tion business...information is the glue that holds together the structure of all businesses' (Evans and Wurster 1997, p. 72). The increasing centrality of information in the production of goods and services, is thus related to the development of information technology. This technology, once wide-scale connectivity has been achieved, permitted the development of an informa-tion economy dealing with information itself, notably as an electronically packaged and transmitted product. The four characteristics proposed by Goddard (1990, 1992), are thus, interrelated.

The information economy consists of five major elements (Figure 1.3b):

- *Infrastructure:* consists of information and communications technologies, including computers, communications appliances, telecommunications networks.

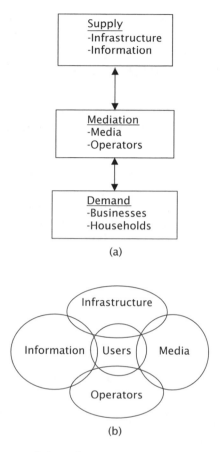

Figure 1.3 Major elements of the information economy

- *Information:* as contents of all types, whether personal, business, educational or entertainment, to be delivered through the infrastructure to customers (users).

- *Media:* through which various types of information are consumed, basically the telephone, TV (and radio), and the Internet, devoted to different information types, at least from the user perspective, and requiring different production, transmission and receiving systems and appliances.

- *Operators:* Consist of a very wide variety of companies that deal with the operation of businesses for production and servicing of the infrastructure (e.g. computer and telephone companies), information (such as news agencies and film-making companies) and the media (such as web design companies and TV stations).

- *Users:* are all customers of the infrastructure, media and information, whether households or businesses.

Traditionally these five major elements amount to three economic functions, namely supply (infrastructure and information), mediation (media and operators), and demand (businesses and households). The supply side of the contemporary information economy is typified by fusion processes among these elements (Figure 1.3b), mainly because all information types have been turned into electronic bits, and also because of the later development of a single channel for all information types, the Internet. The infrastructure, as well as the production and consumption of information will be highlighted in later chapters, so some attention will be given here to the media and operators (Kellerman 1997).

1.3.2 Media

Until the 1970s there existed three separate major information media: TV (and radio) broadcasts which were mostly unwired; the telephone which provided mostly local and inland long-distance conversation services; and computers which functioned as independent data-processing machines (see Kellerman 1993a). The information revolution has brought these three media closer to each other, through improvements in digital technology and transmission channels. Each of these media has been enhanced and expanded as a separate medium as well. The telephone was coupled with fax and mobile telephony, the television was coupled with cable and satellite transmissions, and computers have turned into major appliances for Internet communications. Fibre optics and satellites provided powerful transmission capacities, whereas the integration of the three media permitted visual presentations of computer inputs and outputs on PCs, as well as their transmission from one computer to another. In addition, a wide variety of communications means was developed, mainly digital phone service, fax, the Internet, video conferencing and videophones (see also Castells 2000, Kellerman 1993a).

The integration of transmission media and technologies has several impacts. First, it may become technically possible to transmit all types of information through one transmission system (Batty and Barr 1994). For example, technology permitting the use of standard telephones for Internet-transmitted phone calls already exists. Second, developments in one technology, such as computers, telecommunications or TV, may have direct impacts on other technologies. The pace and significance of technological development are enormous. The exponential decrease in the cost of telecommunications and computers is coupled with an exponential increase in their speed (Makridakis 1995). Third, the international telephone system which, until recently, enjoyed overcapacity, may become overloaded because of the increased use of the Internet system for the transmission of wideband visual information, such as movies (*The Economist*, 1996). Internet providers are considering the introduction of separate lines for web communication, so that other services will be charged differently (Schrag 1996).

Technologically, then, it may become possible to have future information outlets and machines installed in homes and offices, facilities which will serve as terminal points for the transmission of all types of information through a single system, or, from a user's perspective, through a 'one-stop shop'. Such a multi-purpose, unified transmission system is not what Noam (1997) termed an 'integrated single superpipe', related to traditional state monopolistic common carriers, but it is rather what he called system integration, or a 'system of systems', which is to emerge following the present 'network of networks' (i.e. the Internet). System integrators, or the new generation of operators, would provide a variety of telecommunications services to end-users, as well as to competitive suppliers. A similar change is already underway with the introduction of Internet and fax services through mobile telephones.

A similar development was described by David (1990) for the evolution of the electricity industry, moving from the provision of electric power through independent dynamos to central and interconnected power stations. The difference between the central supply of electricity and that of information is that electricity amounts to one product, energy, with endless applications and uses, whereas centralized information supply amounts to an endless number of products served through a single infrastructure and technology.

1.3.3 Operators

Until the 1990s the organizational structure of information production and transmission reflected the technologies of the 1960s, as well as governmental regulative philosophies which called for the existence of separate operators for each type of information. Thus, the provision of telephone, TV, computer and data services was kept separately. Even the breakdown of the giant Bell system in the U.S. in 1993, permitting AT&T to produce computers, did not change this separation. However, several developments in the late 1980s changed this situation. In the U.S., regional Bell companies looked for additional revenues, given the rather restrained local telephone market. The permission granted them to transmit their own cable TV programming over the telephone system was coupled with the development of an opposite technology, enabling the transmission of phone calls over cable-TV systems. These possibilities have yielded several mergers between telephone and cable-TV companies.

Another recent technological development of a potentially much wider application has been the transmission of telephone and videophone calls over the Internet system. This has brought into the operators' game another group of players, namely hardware companies, notably Intel and IBM, which now provide telephone services over the Internet. Thus, there has evolved a blurring of interest and operation areas among previously specialized communications and computer companies (Castells 1996, p. 63, Makridakis 1995). Although the Internet telephone technology is still young, it has gained some popularity, since users pay only for a local call or nothing for accessing their Internet Service Provider (ISP). In the U.S. alone the number of users was estimated at half a million in 1997, and major telephone companies, as well as the Federal Communication Commission (FCC), were exploring the market.

Current trends point, therefore, to a move from a parallel and competitive operation of information companies, each in its traditional speciality, coupled with competition among companies within the same field. Rather, growing competition among companies in areas other than their historical specialization has developed. This competition may bring about changes in the rules of the game. For instance, the sale of information service-packages to end-users, providing telephone service, cable TV and Internet service, or a move from charging per time-use in the account settlement system among telephone companies to charging per width of access, as is the norm for the Internet system (*The Economist* 1996).

In the mid-1990s several major mergers, acquisitions and alliances occurred, topped by the acquisition of Time-Warner by AOL, creating information operators which provide all or almost all information services, namely telephone, TV, computer and the Internet. The converging of hardware and software enables companies to gain control over the entire spectrum of information business from production, through distribution to display (Mosco 1996, pp. 185, 194–195). This has become possible technologically, and in the U.S. also administratively, by changing governmental policies as a result of the 1996 Telecommunications Act, a law which amounts to a major overhaul of the 1934 law which created FCC (Federal Communication Commission) as the federal regulating agency (FCC 1997, Chan-Olmsted 1998, Tseng and Litman 1998). Whereas these mergers and acquisitions may hamper domestic competition and hence reduce service improvements and efficiency, they may simultaneously be balanced through market intrusion by foreign operators. Still the universality of service, traditionally required from telephone operating companies, may be replaced by the so-called *cherry-picking* of service areas (Graham and Marvin 1996, see also Warf 2001).

A central power in this regard is the Internet system which represents a major source of demand and versatility of use: 'The wonderful thing about the Internet is that it still reflects what its users want, not what some large telecommunications company believes they ought to have' (*The Economist* 1996, p. 24). The Internet is also one of a few technologies, jointly with writing, printing, and electricity, which were termed *general purpose technologies* (GPT) (Helpman 1998). It is similar to other GPTs in that it is an enabling key technology, being used simultaneously by many users (Malecki 2000b). However, as far as media and operators are concerned, its major power is the unification of all information forms (text, data, voice, graphics, motion picture) into one information channel.

1.3.4 Information economic geography

Information economic geography focused first mainly on infrastructure, be it computers (Hepworth 1990) or telecommunications (Kellerman 1993a). The implicit assumption was that the infrastructure and its uses are to be priced, among other things, vis-a-vis their location and spatial networking, resulting in spatial patterns of industries and services, at the urban, regional, and national levels. The geographies of infrastructure have proven to be neither decentralized nor concentrated, but rather complex (see Hepworth 1990, p. 97).

Contemporary, mainly Internet based, information economy, assumes foremost that information per se has production, processing, transmission and consumption prices which may vary from one place to another. Thus, some commentators stated that 'information systems redefine and do not eliminate geography' (Li *et al.* 2001, p. 701). Such a redefinition may comprise new networks of centrality and peripherality (see Moss and Townsend 2000, Graham and Marvin 2001). The fusion processes between the major elements of the information economy may imply urban specialization within the information economy which may potentially evolve along one of two lines: phases in the handling of information (production, processing and packaging, transmission, and consumption); and types of information production at large (information, knowledge, innovation). Such local specializations are embedded within the local economy and social structure. If a city specializes in more than one phase, within any of the two classifications, it may become a leader within the information economy. We shall discuss this further in Chapters 3 and 4.

1.4 Information Politics

We have just noted, regarding information economics, that geographers tended to focus on the study of its infrastructure rather than on information per se. A similar trend was noted for information politics. 'The notion that information is a constitutive force in society (e.g. power, control)... has to date not been introduced into geographical studies to any significant extent' (Li 1995, p. 30). Instead, light has been shed on geographical-political issues concerning information infrastructure (or technology), media and operators (see e.g. Warf 2001, Graham 1998).

Information politics relates mainly to the role of the state and the markets in the control and regulation of production and consumption. However, as we shall see, it may have to do with individuals as well. The production and distribution of information is not only dependent on technological infrastructures, but on constant inputs of numerous professionals, who may be located in separate cities, and who produce, edit, transmit and manage the flow of information (Graham 1998). Electronic-virtual information is thus space and place dependent for its production and distribution. The information production industry in its various branches (TV programs and websites as major electronic forms of information) may tend to nest and locate within cities with initial and cumulative expertise in infrastructure production. Such cities may accumulate power in information production, as well as a central, though not necessarily decisive, role in information distribution.

Users may benefit from the growing production of information if they meet several preconditions: they must have technical access to information, mainly to the Internet; they must have socioeconomic access vis-a-vis knowledge to interpret and use the received information; and they must have enough buying power to buy it or get involved in all kinds of e-commerce. Since all these elements depend on previous educational and economic development, information consumption and its uses are not equally distributed in space, and hence the so-called *digital divide*, which we shall discuss further in Chapter 7.

Socially and spatially unequal information production and consumption are developed and sustained by the tendencies of advanced national economies to permit market forces to shape the social geography of information. National governments have gradually retained more minor regulation and control tasks, following four major shifts undertaken since the 1980s (Mosco 1996, pp. 202–203): 1) *commercialization*; changing government policies from public interest (such as universality of service provision) to market standards; 2) *liberalization*, opening telecommunications and information services to competition; 3) *privatization*, selling previously government owned or managed services to commercial companies; and 4) *internationalization*, permitting foreign companies to provide services in domestic markets, either directly or through some alliances with domestic companies.

Interestingly, the role of the state in the production and regulation of knowledge, notably innovative knowledge, has not followed the path of information production. Moreover, the direct and indirect involvement of national governments in this area has increased (see Mitchelson 1999, Kellerman 2002a). Thus, governments provide assistance to universities to develop technological studies, as well as providing budgets for innovative research, the applied results of which may be maintained by the universities or research institutions. Moreover, governments provide seed monies for start-up companies, and they assist or invest in the construction of high-tech industrial parks and industries. Sub-national regional and local governments too are engaged in the formulation and execution of development policies for the attraction of high-tech industries (see e.g. Gibbs and Tanner 1997).

The difference between governmental policies for the production of information and the production and dissemination/application of innovative knowledge is based on the assumption that information currently may be produced best and conveyed most efficiently by the private sector and market forces. On the other hand, however, knowledge production cannot be left to market forces only, because they may keep it too geographically concentrated. We shall elaborate on this in Chapter 3. Another reason for governmental efforts to assist the development of innovative knowledge relates to this knowledge being tacit knowledge, which does not lend itself to instant transmission as information and codified knowledge do. Tacit knowledge, if not available through permanent or temporary migrations of research and development (R&D) workers has to be 'home-grown'. High-tech R&D requires major long-term investments in education and training, as well as incentives for global venture capital to flow and anchor in specific places. Success in the development of high-tech industry in a certain place implies too an accumulation of power there, though in rather different ways than in places specializing in the production of information. As we shall see in Chapter 3, success in high-tech R&D activities may stimulate a mostly local development chain in both research, industrial production, and producer services, with an indirect influence on people's lives outside the region through innovation. On the other hand, the development of a local speciality in information production may have a wider, and even global, impact on people's daily lives through popular television programs or Internet sites.

The development of the Internet and notably the Web, have brought about a shift in the distribution of power not only from governments to the private sector, but also from institutions to individuals (see related discussion in Thu Nguyen and Alexander 1996). The production, and even more so the distribution of information to the public, have traditionally been of the form of *one-to-many*, completely dependent on large-scale and powerful institutions, such as newspapers, radio and TV stations, and book and magazine publishers (see Morris and Ogan 1994, Adams 1998, Dodge and Kitchin 2001, pp. 20l, 56–57). However, the production of a website with either permanent or changing content can be done by any individual, not to mention the varied e-mail systems, ranging from a one-to-one (correspondence) form to *one-to-many*, and *many-to-many* (listservers, etc.). Though the construction of a website still requires a modest investment of capital, it allows extensive populations to produce and distribute information on a global scale. Whereas information production, publication and distribution powers have shifted partially to individuals, they often have relatively little means to market their sites, so that sites may be accessed by just a few surfers, or even none, so that the shift in power of information production may remain more potential than real.

By the same token, access to the Web may potentially permit a limitless exposure to information by individuals and its digestion at low costs, information which is flowing in from all over the globe. Still, however, it was found that websites tend to be linked either to other sites in the same country, or to websites in the U.S., which is the largest information producer (Halavais 2000) (see Chapter 4). Yet another study of ten Asian countries found too that most hyperlinks are to domestic sites, and links to other countries among those studied are foremost to Japan (37%), Singapore (14%), and Hong Kong (11%), probably as the major financial and thus information centers in the region (Ciolek 2002). A shift in the geographical sources of information implies a shift in power, because it brings about more political and cultural transparency than previously available through the local media, but a smart use of these wider sources is still limited by time availability, personal interests and knowledge, as well as culture and language. The Internet was argued to be 'shaped by, and reflect(s), the place-rooted cultures in which it is produced and consumed' (Holloway and Valentine 2001, p. 153). Thus, the Internet which seems to be a seamless global information system, is currently only potentially such, and the geographically restricting forces for limitless information flows are still diverse.

1.5 Information Law

Information law consists of laws which protect some ownership rights of information, and even more so of knowledge and innovative knowledge, known as *intellectual property rights. Copyright* laws have been developed to protect information and knowledge; *licence* requirements have been developed to protect information tools and products; and *patent* laws protect the ownership and rights of innovative knowledge. These legal arrangements may seem to be limiting the free transmission of information and knowledge. However, information laws actually enhance the opposite. They facilitate the transmission of knowledge and information, because without such legal systems for the protection of intellectual rights, information and knowledge transmissions would have been rather more restricted (see Smith and Parr 2000, p. 8–9, Johnson and Lundvall 2001).

The long existing system of information laws has come under threat with the digitization of information and its easy and instant transfer through the Internet. The traditional approach to intellectual property protection has viewed information and knowledge as similar to material products, which can be saved by putting barriers around them for the prevention of leaking, which may cause 'their scarcity value (to be) eroded' (Boisot 1998, p. xiv, see also Barlow 1994). Protection was thus established for the containers (mainly paper products, such as books, documents, etc.), rather than for the ideas themselves.

The Internet challenges several basic elements of law and law enforcement. The Internet blurs the boundaries between the public and the private, as well as between professional publishers or printers and individuals (Elkin-Koren 1996, Geller 1996). The fluid and global nature of the Internet does not fit traditional legal approaches which 'tie the choice of law to points fixed in geographical space' (Geller 1996, p. 28). Still, however, the more knowledge becomes codified the easier it becomes to protect it (Roberts 2000, Boisot 1998). On the other hand, however, the fluid nature of the Internet has brought about intellectual piracy, so that codified knowledge becomes more difficult to hold on to and extract value from, causing some to declare the Internet as a lawless *electronic frontier* (see Warf 2001). At yet another level, the Internet blurs the clear differences between the permitted and forbidden from a spatial-legal perspective. An example in this regard are virtual casinos, operated from territories where gambling is permitted, but players are scattered indifferently in permitted and forbidden areas for gambling (Wilson 2002).

Another legal problem stemming from the geographical fluidity of the Internet is taxation, mainly of sales. This question may receive special attention in upcoming years, once the currently developing so-called *geolocation* projects, which permit the locational identification of computer-hosts participating in any Internet interaction, become mature and reasonably-priced for massive use, in this case by governments (see Chapter 8). The structure of the Internet itself, based on domain names for website identification, requires protection as well (see Wilson 2001a).

Regan (2001) noted the paradox which the Internet poses for the very functioning of legal systems. On the one hand the system seems to be seamless and disembodied, but on the other, it may provide an authentication of place and identity to a much larger degree than activities in the physical world and this authentication of place and identity is crucial for law enforcement. The Internet is also beginning to change the traditional practice of law. The State of Michigan announced the opening of a so-called *cyber court* as of January 2002, in which lawyers not licensed in Michigan will be permitted to practise. In addition, a number of websites operated from various countries, offer international private cyber courts, so that the problem of location-based rules and laws is becoming even more crucial (Wilson *et al.* 2001).

1.6 Conclusion

The preceding discussions pointed to a technological revolution which has rapidly introduced information and communications technologies into the traditional world of information. This rapid introduction and dissemination has required quite radical adjustments of economic, political, and legal systems, at both domestic and international scales, changing the organizational, regulative and legal status of information products, operators of businesses of information production and services, as well as those of consumers. The major question to be treated in the next chapters is, whether geographical adjustments also followed through with these changes. There are those who would argue for the *death of distance* (Cairncross 1997), whereas others would point to the growing importance of a handful of global centers (e.g. Sassen 1991).

The information revolution has yielded contradicting trends in the aspects presented so far. It has brought about huge conglomerates of information production and distribution, but, on the other hand, it has permitted individuals to engage in the production and global distribution of their own information items and products. The flexibility and fluidity of information production and use has not eliminated the need for information and knowledge protection, as intellectual property. Rather it has accentuated the growing dependence on information and knowledge, as its use and dissemination become more flexible.

The contemporary handling of information, looked upon from whichever perspective, implies foremost the application of information technology. Information has become completely integrated with information technology, whether it be for its production, storage, processing, retrieval, transmission, manipulation, or consumption. The immense and constantly growing power of information technology for the handling of information has led to an illusion that information is to be ubiquitously and freely produced and used. It turns out that information is a social and economic product and resource with its own rules of operation, even under the most fluid conditions, manipulated by capital. Thus, information constitutes an integral part of the locally unfolding mosaics of power and culture. On the other hand, the easiness of global flows of information and the virtual image of information turn it into something that may transcend local and domestic conditions. These rather complex relationships will be in the center of our discussion in the next chapter. The different types of communicative materials, especially information, on the one hand, and knowledge and innovative knowledge on the other, may yield varying geographies, which will be the focus of Chapters 3 and 4.

BASICS

2

'Material space and electronic space are increasingly being produced together.'

(Graham 1998, p. 174).

This chapter is devoted to some basics in the study of the geography of information, in particular, electronic information. It will begin with an exposition of the development of the study of information geography, and the conceptual avenue chosen for the following chapters, followed by discussions of the two most basic notions for almost any geographical inquiry, space and place. These notions will be highlighted from the angle of the two relevant issues in the information age: the real and the virtual for space, and the global and the local for place.

2.1 The Scope of Information Geography

The first half of the 1990s witnessed the publication of books on geographical aspects related to the geography of information, all emphasizing information technology, and all published under the series *Belhaven Studies in the Information Economy: Urban and Regional Development*, edited by John Goddard. Thus, Hepworh (1990), followed by Li (1995) studied the geography of computers and information technology at large, coupled with Robins' (1992) collection of articles on various aspects of information technology, with three chapters devoted to geographical aspects of information technology. This line of study was matched at the time by the study of the geographical aspects of telecommunications (Kellerman 1993a, Graham and Marvin 1996). An exception to the accent on technology per se, was Feldman's (1994) *Geography of Innovation*, focusing on the processes leading to the production of information technology, albeit mainly from an empirical perspective.

The introduction of the Internet in the mid-1990s, and its almost immediately becoming a major medium for information production, flow and consumption, have brought about a focus on *cyberspace*. Several recent monographs (Kitchin 1998, Dodge and Kitchin 2001) as well as a reader (Crang *et al.* 1999) were thus devoted to cyberspace. It is important to note in this regard that cyberspace is not synonymous with information. Cyberspace was defined as 'the *conceptual space* within ICTs (information and communication technologies), rather than the technology itself' (Dodge and Kitchin 2001, p. 1). Crang *et al.* (1999) argued for 'the virtual as spatial' (p. 11), and that 'virtuality (then) is not just something which operates through and across space. It is at its heart a spatial phenomenon' (pp. 12–13). Cyberspace is, therefore, an organizing framework and a medium for the handling of information. The phenomenon of cyberspace was mainly dealt with in the above mentioned books as social and cartographic spaces.

Sheppard *et al.* (1999) put forward several interrelated options for geographies of the information society:

(1) *actual geographies* and geographical change in the information age, resulting from the dissemination of information technology;
(2) *virtual geographies* of networks, hubs, etc., as a product of information and communications technologies;
(3) *conceptual geographies* of areal surfaces within people's minds, when exposed to their description via information technologies, notably geographical information systems (GIS).

The approach chosen here for the following chapters is mostly an *actual geography*, but one that is foremost *of* information per se, accompanied by the study of geographical aspects of information interwoven into urban and economic geographies at large. The study of the actual geography of information is and must be accompanied by inquiries into the close interrelationships between the actual geography *of* information and virtual geography, as a geography of networks.

Information is viewed as an abstract object, rather than a phenomenon or space. As an object it has its own geography. Cyberspace constitutes a phenomenon of virtual space, consisting of Internet information and its users. Information, on the other hand, enjoys geographical dimensions, spaces and aspects beyond cyberspace. Electronic information is produced in real places, and used in real places as well. It is not merely produced and used in electronic forms, but in printed forms as well. Electronic information is transmitted globally through networks of real cables, satellites and hubs. Some of it may be displayed through local real facilities, such as cinemas. Information is an economic as well as a social resource and product, produced and manipulated in a similar way to other resources and products, as far as the roles of capital and social relations are concerned. It can be produced and exchanged even if it has no commercial value. It can be sold and transmitted in various ways, and all these aspects have their own geographies. By its very nature, information may be interpreted and manipulated. This seemingly non-spatial dimension of information has received a spatial accent with the emergence of the World Wide Web, in which geographical language has become a major tool for the structuring, organization and use of cyberspace (Kellerman 1999b).

Moreover, the use of information involves personal and social experiences, which have been enhanced through the Internet as cyberspace. Cyberspace has brought about the emergence of virtual communities (e.g. the San Francisco-based WELL), leading to possible conflicts between the real and the virtual, as well as between global and local communities. These will be highlighted through the following discussions of the meanings of space and place.

2.2 Space

Space may be considered one of the primal notions of geography (see Kellerman 1989); its constitution, meaning, and expression being enormously varied (see e.g., Simonsen 1996). Space as a dimension of wide geographical extent frequently carries either a physical or an abstract connotation (see Gregory *et al.* 1994 p. 8), but it has social meanings as well, some of which are of particular importance for information geography, whether in real or virtual space. We will therefore focus our attention on social space.

2.2.1 Social space

Social space constitutes a relational rather than an abstract dimension, so that it receives its meaning from social relations rather than from its being an object (Graham 1997). Social space has received a large variety of attributes, interpretations, and metaphors (Figure 2.1). It has been claimed to constitute

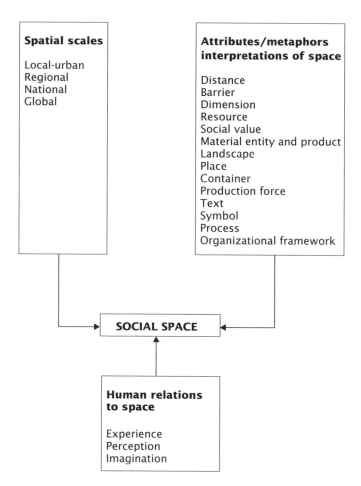

Figure 2.1 Dimensions of social space

distance (Ilchman 1970), barrier (Harvey 1989), dimension (Ullman 1974, Massey 1992), resource (Kellerman 1989), social value (Sack 1980), material entity and product (Lefebvre 1991), landscape (Cosgrove 1984), place (Giddens 1990, p. 18, Entrikin 1991, Merrifield 1993), container (Kellerman 1989), production force (Swyngedouw 1992a), text (Barthes 1972, Duncan and Duncan 1988), symbol (Lefebvre 1991), process (Merrifield 1993), and organizational framework (Soja 1989).

The different meanings and aspects of space are expressed, mediated, or embodied along several geographical and social scales, namely the local-urban, regional, national and global (see Massey 1992). These spatial scales do not merely constitute a convenient professional, administrative or political classification, as human spatial practices and experiences may be perceived and organized along these scales (see Shamai and Kellerman 1985). Typically, space has been treated through the perspective of one 'fixed' such scale, chosen for a specific study, 'so rather than relating scales to one another, spatial effects are regarded as the product of one scale with other scales at best viewed as residual or emergent' (Agnew 1993, p. 252).

A third sociospatial classification focuses on social relation to space, differentiating between material spatial practices, representations of space, and the spaces of representations (Lefebvre 1991, see also Merrifield 1993 and Kirsch 1995). These relations were interpreted by Harvey (1989, pp. 220–221) as experience, perception, and imagination, respectively. Harvey provided further detail to the spatial expressions of these relations under varying spatial practices: accessibility and distanciation; appropriation and use; domination and control; and production of space.

Social space constitutes multiple junctions between any of the attribute(s) of space, at any geographical scale(s), related to or mediated through human relations. For example, urban space may be considered a text for the design of a website in virtual space, as is the case in various *MOO's*, and thus it constitutes *imagined space*. On the other hand, urban space may be considered a resource for the real location of an information business of a webmaster who designs this website. Social space constitutes, therefore, many things and may carry several meanings even at the same time and place, and more so at different spatial scales and for different groups of people or individuals. These many meanings of space apply not only to spatial practices, but even more to students of space. Space has been used extensively and sometimes ambiguously in social theory writings, notably in discussions of the power of space and its influence, as well as in discourses concerning its status relative to time. Castells (2000, p. 441) argued recently that 'space is not a reflection of society, it is its expression. In other words: space is not a photocopy of society, it is society'.

2.2.2 *The real and the virtual*

Virtual space or *cyberspace* may be considered a special form of social space in several of its attributes: resource and production force for e-commerce; a text and a symbol by its very nature; and as an organizational framework in almost all of its uses and applications. Some would argue for its being a landscape, a place and even a social value (see Dodge and Kitchin 2001 for detailed discussions). It is by far an *imagined space* of representation through its virtual imitation or virtual description of real space and place. Its geographical scale is most complex. On the one hand it spans globally, but on the other hand it may visually present local and regional images and information as well. It constitutes a 'different human experience of dwelling in the world; new articulations of near and far, present and absent, body and technology, self and environment' (Crang *et al.* 1999; p. 1).

Virtual space and real space are different, but still highly interrelated, and in various ways. The following discussion will highlight these differences and interrelationships. Cyberspace is distinguished from real space in many instances, which may be divided into three groups of dimensions: organization, movement, and users (Table 2.1). Li *et al.* (2001) believe that speed of movement is the most important dividing dimension between the two forms of space, despite restraints on the speedy movement of information, such as infrastructure. This is followed by content, which distinguishes cyberspace from real space in that it is devoted to information only. The fluidity of production and consumption of cyberspace should be added as a third most important difference between real space and cyberspace. It is expressed in the ease of construction and demolishing of sites in cyberspace compared to construction activity in real space; the undefined identity of users in many cases in cyberspace, and the lack of commitment to any given 'place', 'site' or 'space' in cyberspace, coupled with the rather imaginative spatial experience it provides. Dodge and Kitchin (2001, p. 16) suggested, in this regard, the possible interpretation of this fluidity in light of Relph's (1976) notion of *placelessness*, proposed at the time for the sameness of real landscapes, mostly in urban areas.

Yet another dimension; distance, received much publicity through the provocatively entitled book *The Death of Distance* (Cairncross 1997). The decline in the importance of distance has been noted by several additional writers concerning telecommunications technologies at large (e.g. Gillespie and Williams 1988, Gillespie and Robins 1989, Atkinson 1998, Brunn and Leinbach 1991, Castells 1989 and Negroponte 1995), but the partial decline in the significance of distance for informational and transactional interactions should not be viewed as signalling a decline in the importance of real space altogether (as proposed by Benedikt 1991). The low or complete lack of importance of distance for virtual interaction does not nullify a related barrier typifying interaction in real space, namely access. In both

Table 2.1 Real and virtual spaces

Dimension	Real space	Virtual space
Organization		
1. Content	Physical and informational	Informational
2. Places	Separated	Converge with local real ones
3. Form	Abstract or real	Relational
4. Size	Limited	Unlimited
5. Construction and maintenance	Expensive and heavily controlled	Reasonably priced and lightly controlled
6. Space	Territory/Euclidean	Network/logical
7. Matter	Material/tangible	Immaterial/intangible
Movement		
8. Medium	Transportation	Telecommunications
9. Speed	Depends on the mode of transport	Speed of light, constrained by infrastructure, costs, regulations, etc.
10. Distance	Major constraint	Does not matter mostly
11. Time	Matters	Matters, but events can suspend in time
12. Orientation	Matters	Does not matter
Users		
13. Identity	Defined	Independent of identity in real space
14. Experience	Bodily	Imaginative and metaphorical
15. Interaction	Embodied	Disembodied
16. Attitude	Long-term commitment	Uncommitted
17. Language	National-domestic	Mainly English-international

Source: Items 1–2, 8–11, 12 (Li *et al.* 2001); items 3, 14–15 (Dodge and Kitchin 2001, pp. 30, 53); items 6–7 (Graham, 1997).

spaces, barrier factors for access are of a similar nature: infrastructure limits (vehicle/computer power; road/bandwidth; socioeconomic constraints, etc.). The results are also similar: inefficient use by those who can, and lack of participation by those who cannot. Time has been considered by some commentators to supercede location as the most crucial and basic dimension, since many sources of electronic information compete at a given time on a user's attention (see Dodge and Kitchin 2001, p. 14).

Besides these differences, real space and cyberspace are interrelated in many ways. On the one hand, cyberspace is dependent on real space for its infrastructure (computers, telecommunications, offices, etc.), or it requires some *spatial fixity*. This infrastructure is not equally distributed in real space, and hence a real geography of the virtual one (Malecki and Gorman 2001, Dodge and Kitchin 2001, pp. 14–15). On the other hand, real social space is not independent of virtual space any more. It could be argued, until the introduction of information technology, that material space was relatively independent of imagined space, followed by some complementarity between real space and early cyberspace. It seems that the contemporary tremendous integration of information technology and the Internet into all spheres of economic and social lives, makes it now impossible to manage and manipulate real social space without the use of, or reference to, virtual space. This relates not only to smart buildings replacing conventional ones (see Batty 1997), but to the maintenance of conventional real estate as well. By the same token, production and consumption of material products and services in real space are interwoven with Internet activities, whether it be the obtaining of relevant information, the performance of business correspondence, maintenance assistance, or consumption through cyberspace only (e-commerce). The production and consumption of non-electronic information, such as books, is also interwoven with virtual space, through computerized editing, for example, at the production end, and browsing and e-shopping at the consumption end.

The exact nature of the relationship between real and virtual spaces has been discussed and commented on by numerous commentators. Thus, Batty (1997, p. 341) suggests that 'space and place has been influenced by the gathering momentum of the digital world', and he further referred to 'the impact of computers and communications on place itself'. Other commentators have noted more complex patterns of interrelationship. Thus, 'cyberspace is hardly immaterial in that it is very much an embodied space' (Dodge 2001a, p. 1), and from the other end, 'information systems redefine and do not eliminate geography', and 'electronic space is embedded in, and often intertwines with, the physical space and place' (Li *et al.* 2001, p. 701). Thus, the Internet 'is shaped by, and reflects, the place-rooted cultures in which it is produced and consumed' (Holloway and Valentine 2001, p. 153).

Graham (1998) went one step further than the previous formulations for the absolute or relational views of real and virtual spaces, by suggesting that both spaces co-evolve in that they 'stand in a state of *recursive interaction*, shaping *each other* in complex ways' (p. 174). This two-way process may bring about a 'liberation' of activities previously taking place in real places, but their moving into cyberspace creates new real and fixed locations for the functioning of cyberspace (see Swyngedouw 1993).

Summing up the numerous observations so far, there are several interrelationships between real and virtual spaces:

(1) *interdependence* in their very functioning;
(2) *co-evolution* of both spaces;
(3) *dual construction and elimination* of sites and activities in both spaces.

The interconnection and co-evolution of real and virtual spaces is of special significance when it comes to the geography of information. Information, the only content of cyberspace, is produced in real space by real people, but its very production and consumption are dependent and are embedded in virtual space, the major contemporary medium for its production, transmission and consumption. The construction of new websites may bring about the development of a real world industry, and vice versa: the very existence of such an industry and expertise, may bring about further developments in cyberspace. At the same time, the availability of information services on cyberspace, handled from specific centralized real world places, may bring about the elimination of local facilities (e.g. local travel agencies replaced by Internet travel services).

The complex interrelationships between the real and the virtual, notably the mediation between the two, may be illuminated by a conceptual development of the Lefebvre–Harvey approach for the production of space (Lefebvre 1974, Harvey 1989, pp. 218–222), mentioned before. Figure 2.2 presents both virtual and real spaces as elements of the more general social space. Material spatial practices constitute real space flows, transfers, and interactions. Virtual space is presented as a space of imagination. The two are mediated through codes (information) and knowledge, presented as representations of space. The construction of virtual space, thus, is embedded in real space through a double experience: an *operational* one relating to the possible transfer of activities from real space to virtual space, or their improvement through complementarity in virtual space, coupled with a *metaphorical* experience, through reflection, leading to the use of geographical language, symbols and tools in the construction and use of cyberspace. The cumulative experience in using virtual space leads, again through human knowledge and codes, to a reshaping of real space, in terms of locational patterns of both production and consumption.

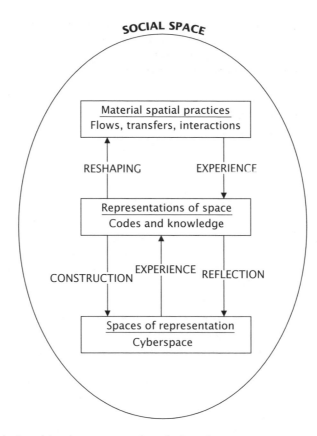

Figure 2.2 Relationships between real and virtual spaces

The constant discourse between real (material) and virtual (imagined) spaces through knowledge and information (perception) accentuates the oneness of the material, the perceived and the imagined, as far as social space is concerned. Harvey (1989 p. 219) noted such an option, when he stated: 'The spaces of representation, therefore, have the potential not only to affect representation of space but also to act as a material productive force with respect to spatial practices'. Information and knowledge constitute not only objects of production and consumption, in both real and virtual spaces, but they also constitute mediating forces between the construction and reshaping of real and virtual spaces.

2.3   Place

2.3.1   Interpretations of place

Place 'has to be one of the most multi-layered and multi-purpose words in our language' (Harvey 1993, p. 4). Its geographical daily meaning is defined by *The Oxford English Dictionary* as 'a portion of space in which people dwell together' (Simpson and Weiner 1989). Along this tradition, Castells (2000, p. 453) recently defined place as 'a locale whose form, function, and meaning are self-contained within the boundaries of physical contingency'. One may view these basic definitions of place as a bottom layer in a multilayer identity or interpretation of places emerging in a globalized and cyber-based world (Figure 2.3). In the following paragraphs we briefly examine these layers of interpretation.

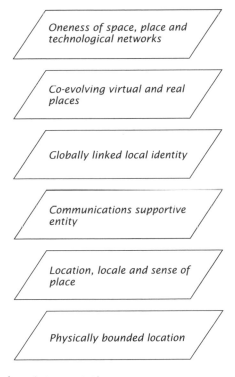

Figure 2.3 Layers of place interpretations

The notion and connotation of place in geography has been transformed from viewing it as a bounded location distinguished only by its physical character, 'to the conceptual fusion of space and experience that give areas of the earth's surface a "wholeness" or an "individuality"' (Entrikin 1991, p. 6). Thus, places involve three complementary elements: location, locale, and sense of place (Agnew and Duncan 1989, Agnew 1993). The interrelationship between space and place may constitute more than physical context, as 'space provides the context for places but derives its meaning from particular places' (Relph 1976, p. 8).

Place is not only a physical entity loaded with a cultural/human identity and interrelated with space. It may be further defined by its 'physical and social structures that support communications'. Thus, 'place is an interweaving of communication and action' (Adams 1998, p. 94). Adams (1998) further noted that places used to function through communications networks which involved physical entities, such as barriers and bodies, whereas computers present communications networks which consist of signal flows and codes, thus bringing about more complex place structures and identities.

In a globalized world, places may turn into 'traces of movement, speed and circulation' (Thrift 1996, p. 289), with both local and external sources for idiosyncratic place identities. Pred (1984) and Pred and Watts (1992, p. 11) have emphasized the decisive role of local social relations and events, even when local activities are decided upon nonlocally. On the other hand, Giddens (1990, pp. 18–19), Massey (1994) and Gupta and Ferguson (1992) have all accentuated *external* social and economic relations as shaping constantly dynamic places. Thus,

> *people are everywhere conceptualizing and acting on different spatialities ('A global sense of place').... And the particularity of place is...constructed not by placing boundaries around it and defining its identity through counterposition to the other which lies beyond, but precisely (in part) through its specificity of the mix of links and interconnections to that 'beyond'. Places viewed this way are open and porous.*
>
> *(Massey 1994, pp. 4–5).*

Places are not only receivers of external inputs and local processors of these inputs jointly with local ingredients; they serve further as producers of electronic information and virtual places, embedded to some degree within local social cultures. These local cultures are more weakly expressed when a virtual place serves as a meeting point for people located in different physical places and countries (see Li *et al.* 2001). Domestic languages are replaced by English, and the virtual meeting place lacks any cultural identity or historical heritage. The global-virtual meeting point may either be used on a permanent basis, or it may dissolve following a single or several meeting sessions, but its meeting parties remain each in their real culture-loaded places.

The growing use of cyberspace as a place for various activities, has brought some commentators to predict 'that defining space and place separately from technological networks soon becomes as impossible as defining technological networks separately from space and place' (Graham 1998, p. 181). Place has thus become a complex entity, defined simultaneously in real and virtual spaces, serving as loci of production, consumption, interaction, and identity.

2.3.2 *Globalization*

Globalization has been extensively studied over the past two decades, notably following the World City Hypothesis originally presented by Friedmann and Wolff (1982) (for a recent review see Sassen 2001a). The study of globalization focused to a large degree on global cities (New York, London, and Tokyo) (see e.g. Sassen 1991), and various lower level world cities (see e.g. Short and Kim 1999). One of the attributes of globalization and the formation of world and global cities has been the introduction of information technology and the resulting facilitation for almost limitless worldwide flows of information. The information revolution was coupled with the capital revolution permitting instant global flows of capital, following the removal of state restrictions, and the growth in multinational corporations (MNCs). Thus, capital turned into digital information: 'nowadays money is essentially information' (Thrift 1995, p. 27).

Appadurai (1990) distinguished between five dimensions of global cultural flows, all of which were termed 'scapes', since they are looked upon differently by various actors, and since they are fluid and irregular. The five were: *ethnoscapes* (the migration of workers); *mediascapes* (television, movies, magazines, etc.); *technoscapes* (technology); *finanscapes* (finances); and *ideoscapes* (ideologies). To this list one can add *commodiscapes*, for commodities which carry some cultural messages on a global scale, such as pizza as an Italian food, or Japanese cars (see also Knox 1995, p. 245), and *infoscapes*, for information and knowledge globally transmitted mainly via the Internet, if not included in mediascapes. All these flows are asymmetrical and they reflect the domination of world cores (see Castells 1994).

41

Viewed from a political economy perspective, Mosco (1996, pp. 205–206) considered the spatial agglomeration of capital as the source of globalization, which has led to a changing geography of information and communications, in which previously dominating major centers, such as New York and London, have enhanced their economic power, through an extension of their area of influence (distanciation). Even within these global centers, the growth in economic power was not equally distributed throughout their metropolitan areas. Giddens (1990) interprets globalization as a rather social process, constituting an integral part of the local experience: 'globalization can (thus) be defined as the intensification of worldwide social relations which link distant localities in such a way that local happenings are shaped by events occurring many miles away and vice versa' (p. 64).

Another perspective on global flows and location was put forward by Castells (1989), namely the dissolution of local space, coupled with the emergence of the *space of flows*. The space of flows was defined as 'the material organization of time-sharing social practices that work through flows' (Castells 2000, p. 442), where flows include all possible ones except for people; namely: capital, information, technology, organizational interaction, images, sounds, and symbols. It should be noted, however, that information flows (including those of sounds, images and symbols) have become an integral part of the direct and everyday lives of people in developed countries. On the other hand, the global flows of capital carry more indirect impacts for most people, and the flows of technology may have rather incremental leap impacts on people's lives rather than continuous daily ones. We shall return to this point later on.

The space of flows, as developed by Castells (2000, pp. 448), consists of three layers: the first one constitutes a *circuit of electronic exchanges* embodied in networked cities; the second is a layer of *nodes and hubs*, hierarchically organized and topped by global cities, which serve as major loci of information production; the third layer consists of the *managerial elites*, charged with the directional functions of the space of flows. We noted already that one of the most significant aspects of the globalization of information flows is the active participation of people from all over the world, as both producers and consumers of information. This widely used access does not reduce the importance of the managerial elites: 'elites are cosmopolitan, people are local' (Castells 2000, p. 446). However, it differentiates the roles of actors in the very constitution of global flows and their integration into their daily lives.

The exact relationships between the rather global space of flows and the domestic spaces of places are not made clear in Castells' writings. In one place, Castells (1985, p. 14) argues for 'a space of flows substituting a space of places', and in another place he is referring to 'the historical emergence of the space of flows, superseding the meaning of the space of places' (1989, p. 348). Elsewhere Castells attributed the supremacy of the space of flows more specifically to organizations in the informational economy (1989, p. 169). The very idea of the superiority of the space of flows was criticized, with the observation that the locational anchoring of the space of flows within the space of places, and the dialectic relationship between location and flows were ignored (Merrifield 1993, Kellerman 1993a). We shall return to the relationships between the spaces of flows and the spaces of places in our discussion of the digital divide (Chapter 7).

2.3.3   The global and the local

The *global* and the *local*, *space* and *place*, together and separately, have long been foci of research and debate in geography (for recent reviews see Yeoh 1999, Unwin 2000), in sociology (see Eade 1997), as well as in political science (see Hewson and Sinclair 1999), and urban theory (see Smith 2000). Some commentators went as far as stating that 'the local and the global are mutually constituted' (Swyngedouw 1997, p. 137). It would suffice here to note that definitions and perspectives on these notions are neither simple nor clear, nor generally agreed upon. Some writers would argue that the contemporary local in a globalizing world, or 'the traditional fixed statics of space are becoming eclipsed by a new fluid dynamics of pace' (Luke and Tuathail 1998, p. 72), and 'places are conceptualized in terms of their ability to accelerate or hinder the exchanges of global formations' (ibid., p. 76). On the other hand, others would argue that:

> the 'global' in the dominant discourse is the political space in which a particular dominant local seeks global control, and frees itself of local, national and international restraints. The global does not represent the universal human interest, it represents a particular local and parochial interest which has been globalized through the scope of its reach. The seven most powerful countries, the G-7, dictate global affairs, but the interests that guide them remain narrow, local and parochial.
>
> (Shiva 1993, pp. 149–150).

Global flows are complex in their interrelationships with local and even personal life and realities. These interrelationships have been termed *glocalisation* (Swyngedouw 1992b; Robertson 1995), referring among other things, to the personal construction of various localities through global flows of information. The direct interplay between the local and the global implies a declining importance of the national dimension, or governmental control and intervention. 'Globalization (thus) represents a redefinition of places as juxtapositions of intersecting, overlapping, and unconnected global flows and historical fixities' (Amin and Thrift 1994, p. 10).

The interrelationships between the local and the global may dialectically constitute and produce three different conditions or processes represented at the local level: *disembedding, phantasmagoria*, and *fusion*. The first process, disembedding, relates to a possible separation between the local and the global. The second process, phantasmagoria, refers to concealed relationships. The third one, fusion, presents an integration between the global and the local. The three processes may potentially evolve as three stages in on-going relationships between the global and the local, or they may present themselves as independent processes. As Figure 2.4 shows, in each of these processes there are four participating dimensions and forces: *space* (or the global), *place* (or the local), *push* elements from the local to the global, and *pull* elements from the global to the local.

The first possible interrelationship between the local and the global is *disembedding*. This was outlined by Gregory (1994, p. 121), based on Giddens (1990) and Harvey (1989): 'disembedding mechanisms separate interaction from the particularities of locales' (Giddens 1991, p. 20). For example, social life and identities in a given society may be disembedded from their local and national traditions and histories through the intrusion of global media, such as cable TV. Robins and Cornford (1994, pp. 220–221) termed such disembedding processes *deterritorialization* (see also Tuathail and Luke 1994). On the other hand, however, cultural importation of values and identities through global media may yield new forms of communal cohesion.

Disembedding may be brought about by two processes which have been identified for the emerging global/local relations: *time–space distanciation*, a *push* or separation process, and *time–space compression*, a *pull* or cohesion process (Figure 2.4a). Operating simultaneously, they amount to a 'local-global dialectic' (Gregory 1994, p. 118). *Distanciation* is defined here as the geographical separation between local place and global space under intensifying relations between the two. Disembedded electronic space may, for example, be banking or a stock-market investment transaction performed

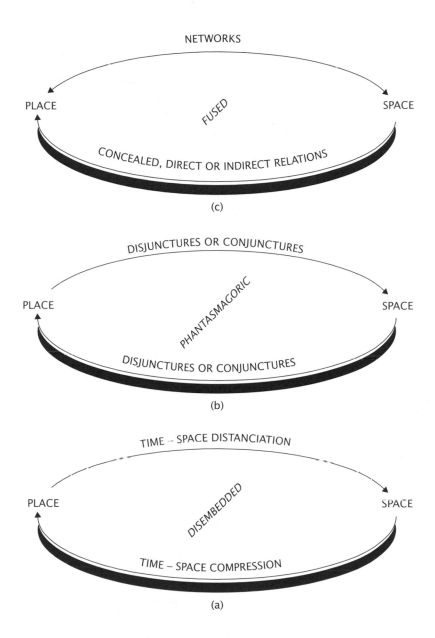

Figure 2.4 Processes between space and place ((a) after Gregory 1994, p. 121)

in New York by an order of a British customer investing in a Japanese firm. The disembedding here relates to global interaction using electronic media for connections from New York to Britain and Japan, seemingly disembedded from local New York life. However, such disembedded electronic space may yield embedded spaces or places, given the frequent need for face-to-face meetings in New York, in order to coordinate and facilitate complex global investments. Such meetings may be typical to such transactions, notably in major global cities (New York, London, and Tokyo) (Thrift 1996, pp. 231–232). One may view these two trends of disembedding and embedding as mutually reinforcing. Disembedded space requires concrete meeting places for coordination, decision-making and information-sharing. These, in turn, enhance disembedded investment space and vice versa.

> *In premodern societies, space and place largely coincided, since the spatial dimensions of social life are, for most of the population...dominated by 'presence' – by localised activity... Modernity increasingly tears space away from place by fostering relations between 'absent' others, locationally distant from any given situation of face-to-face interaction. In conditions of modernity...locales are thoroughly penetrated by and shaped in terms of social influences quite distant from them.*
>
> *(Giddens 1990, p. 18).*

Time–space compression refers to the contemporary '*compression* of our spatial and temporal worlds' (Harvey 1989, p. 240), or a *pull* mechanism, induced by contemporary telecommunications. For example, a telephone call taking place between Australia and the U.K. implies that one of the parties may be awake late at night or working at that time, so that both time and space differences have been compressed. Time–space compression is thus both an outcome and a cause for distanciation, or a separation between the local and the global, hence the local–global dialectic. Time–space compression is not synonymous with Janelle's (1968, 1991) *time–space convergence*, since compression relates to conditions of social space, whereas convergence is a measured index, defined as 'the rate at which places move closer together or further away in travel or communication time' (Janelle 1991, p. 49). Time–space compression reflects power relations. There are those who sense this compression only passively or indirectly, whereas others are in charge of this compression through their handling of the local/global transfers, notably those of capital and information (Massey 1994, p. 149).

The second possible condition or process between the local and the global is *phantasmagoria*, or concealed relations. Phantasmagoria in this regard was briefly outlined in the writings of Giddens (1990, p. 19) and Appadurai (1990), referring to local invisible conditions which reflect those at other, rather distant, locations. For instance, local demand for labor in the Philippine textile industry may result in changing social and spatial patterns, because of rising incomes and migration into centers of textile production. It may further lead to the development of local centers for the recruitment of manpower for overseas employment (Tyner 2000). However, this rising demand for manpower is determined by remote decision-making and capital sources of multinational corporations headquartered in the U.S. What happens in one specific locality may yield similar or contradictory patterns and processes in other places worldwide. In other words, rising demand for textile products in the U.S. may result in different social patterns in newly industrializing countries in Asia or South America. The international interaction in our example of the U.S. and the Philippines, involves not only information exchanges of an economic nature, but cultural ones as well. The global imprints on the domestic scene are not necessarily directly revealed, nor must they be visible, hence the phantasmagoria, rather than a 'separate' disembedded activity of a 'foreign sector'. Under circumstances of globalization the phantasmagoric relationships between place and space may reflect disjunctures or conjunctures (see also Appadurai 1990) (Figure 2.4b). 'The "visible form" of the locale conceals the distanciated relations which determine its nature' (Giddens 1990, p. 19).

The third possible process between space and place under global flows is *fusion*. Here space constitutes the global, defined neither as other specific places nor as an abstract dimension, but rather as flows of elements (capital, information) gathered from many individual locations through networks which turn the numerous local ingredients into global movements (see Goddard 1995). A full cycle of fusion implies the existence of geographical origins, movements, or transmissions, and geographical destinations. In other words, one or several local factors become global, so that they can be moved elsewhere and fused with local space. For example, capital is gathered by a New Yorker investment fund from various European and Asian investors, and then invested through the New York Stock Exchange in a South American industrial plant. Or, as yet another example, information provided by the Internet users in various countries is put on a website managed and hosted in one country, and then transmitted to a global audience.

47

Fusion is, therefore, a double process: the production of global movements out of local ingredients, and the integration of the local with the global at a specific destination locality. The local scene may then express the global in varied patterns, concealed as well as revealed, direct or indirect (Figure 2.4c). For example, the South American industrial plant in our example may be constructed to fit with local patterns of design but managed in a North American way, or vice versa. Another example may relate to the fusion of imported Internet information into domestic life in varied aspects, ranging from human rights to fashion and entertainment. The idea of fusion between the local and the global has been briefly mentioned by Pred and Watts (1992, pp. 5–6), and Massey (1994, p. 5), who pointed explicitly to the evolving strong bonds between the local and the global, even to the degree of partial mutual constitution between them. The various local processes of adoption of, or resistance to, the global, may lead to what Knox (1995, p. 246) termed *reterritorialization* of places.

Table 2.2 summarizes the three processes between the local and the global outlined above. Each of the three processes has its own characteristics. However, they may also be alternatively viewed as phases in a process of evolving relations between the local and the global. Such a process may start with the expansion of the extent of local relations and a possible separation between the global and the local (disembedding). At a second phase, the relations between the local and the global may be typified by specific imprints of other places on a specific locality (phantasmagoria), or alternatively, such relations may begin with these impacts. The evolving relations between the local and the global may reach a third (or second) level through an integration of the local with the global (fusion). Such an evolutionary two or three phase process may emerge in cities/countries undergoing extensive industrialization processes. One may expect especially that concealed global impacts in early phases of industrialization may mature, at a later stage,

Table 2.2 Processes between the local and the global

Process	Interpretation of space	Nature of relations
Disembedding	Abstract dimension	Expanded geographical extent
Phantasmagoria	Specific distant places	Revealed or concealed Disjuncture or conjuncture
Fusion	Global flows	Global networking and local integration

into fusing relations. On the other hand, in developed economies/cities, the fusion of the global with the local may possibly reduce disembeddiment effects, by the very nature of the fusion process.

2.3.4 Place production

'The contradictory experience of being somewhere and nowhere at the same time is perhaps the most obvious cognitive dissonance resulting from the use of the WWW' (Kwan 2001a, p. 26). The emergence of global social networks, and the growing interaction with global information networks bears upon the *sense of place* of users and on processes of *place production*. Halbwachs (1980, p. 134) proposed the terms *implacement* and *displacement* for social reactions to urban changes. By the same token, the simultaneous sensing of local–physical and global–virtual places may be termed *coplacement*, or *copresence* (Shields 1996). The major aspect in this regard is the growing tension between the distant and the local, the absent and the present, or sometimes between disembedded space and place, expressed in distanci-ation and time–space compression (see Giddens 1990, p. 118; Bird *et al.* 1993). Distanciating transportation and communications may bring about a so-called *hyperspace*, referring to experienced space which does not coincide with the place where it occurred. Conflict may thus arise between imported images of places and a possible, later, lived experience in them (see Gupta and Ferguson 1992). By the same token, when the experienced virtual space is metropolitan and the experiencing people live in real nonmetropolitan areas, then the distance between the two may produce fantastic *imagined worlds* (Appadurai 1990).

The development of a sense of place around real places is a long-term and continuous process; which is not the case for instantly replaceable virtual places. These *spaceless places* (Ogden 1994, p. 715) do not provide the physical sensing of places, nor a third dimension of depth, natural movements, air breezes and winds, or smells and sunshine. In addition, virtual places have no history and may not be embedded in a collective memory. The exposure to distant places and people may thus bring about the need to strengthen local identities through the fostering of local heritage. However, the involvement of mass media in processes of reconstruction of local identities may turn such a trend into a rather synthetic *'global' localism* (Thrift 1994b, see also Massey 1994, Castells 1994).

49

2.4 Conclusion

In his critique of contemporary urban social theory, Smith (2000, p. 157) argued: 'The global–local duality in social theory rests on a false opposition that equates the *local* with a cultural space of stasis, ontological meaning, and personal identity (i.e. the "place"), and the *global* as the site of dynamic change, the decentering of meaning, and the fragmentation/homogenization of culture (i.e. the "space" of global capitalism)'. Instead, he suggested that 'everyday life is neither a fixed spatial scale nor a guaranteed site of local resistance to more global modes of domination, whether capitalist or otherwise. Rather, under conditions of transnational urbanism our everyday life–world is one in which "competing discourses and interpretations of reality are already folded into the reality we are seeking to grasp" (Campbell 1996, p. 23)' (p. 118). Castells (2000, p. 446) believes that a possible cultural domination of the global over the local is determined by the following: 'the more a social organization is based upon ahistorical flows, superseding the logic of any specific place, the more the logic of global power escapes the socio-political control of historically specific local/national societies'.

The contemporary world is thus typified by rather fluid identities, constantly exposed to a power struggle between local cultures and information, on the one hand, and global, virtually and economically transmitted ones, on the other. Viewing the local and the global, and the real and the virtual, from the standpoint of people as human agents, rather than from the perspective of material or virtual geographical entities, then the basis for grasping and functioning of human beings is their knowledge and the information which reaches them, and which is interpreted by them for use in their daily lives. The most basic contemporary change in the everyday lives of people in developed countries is the constant exposure to information reaching them from local and global sources, through the same media and for the same costs. The driving power behind the generation of these geographically varied flows of information might be capital and capitalism, but for urbanites' everyday lives it is information that is the point. The receivers of these global and local, real and virtual informations are simultaneously also producers of information which may be transmitted through the same click on the computer keyboard either to a friend or a business across the street, or to a global distribution.

Information systems and information and knowledge per se serve, therefore, as crucial factors for the very existence of the virtual and the real, the global and the local, but they further constitute mediators between these worlds. Simultaneity seems to be the name of the game. We experience simultaneous and on-going operations of people, as *passive* receivers and consumers, on the one hand, constituting at the same time also *active* producers of various types: producers of information; seekers of information on the Internet; knowledge producers; and economic actors, all within complex dimensions of space and place.

TECHNOLOGY

3

'The production of technologies (i.e. knowledge) and know-how are not becoming placeless.'

(Storper 2000b, p. 47).

3.1 Information and Technology

The word 'and' inserted between the words information and technology in the title of this section may seem a bit odd, given the almost trivially used term *information technology*. Information technology is actually the wizard behind the massive development in all the phases of information handling: production, transmission, dissemination and consumption. It is important, however, to examine the relationships between information and technology, because they are of a double nature. On the one hand, information technology facilitates the enormous and complex handling of information through its transformation into electronic bits. On the other hand, however, the massive development and production of the technology itself, in both its hardware and software components, requires intensive knowledge, and foremost innovation. Innovation has become crucial with the ever shortening of product life-cycles (Malecki 2000a).

This distinction between the two relationships between technology and information at large is associated with different geographies. The specific geographies of information production, transmission and consumption, facilitated by information technology, will be dealt with in detail in the next chapters, so that only a rather general discussion on information and technology in this regard will be provided in the next section. The majority of this chapter is devoted to knowledge and innovation which lead to the development and production of information technology and to their geographical behavior and patterns. As such, this chapter is not only on technology, but to a large extent on knowledge and innovation as information classes.

The two geographies, the one of information per se, and that of knowledge and innovation are different in their basic nature. As we shall see in this and in the following chapters, information production is not equally distributed among and within cities in developed countries. The industry of electronic information production is dependent on telecommunications infrastructure, on multimedia, *and* on the high-tech industry. Even the hidden assumption that the availability of sufficient transmission systems will permit equal consumption patterns of information, will prove false. On the other hand, the geography of innovation, in its locational and flow aspects, mainly for information technology, is based on scientific and technological tacit knowledge, patenting protection, venture capital, sophisticated producer services, and entrepreneurship. It will be shown that the geography of innovation is much more concentrated in its geographical behavior and patterns than the geography of information production. Cities which are strongly present on both the geographies of information and innovation will enjoy a leadership status in the wider spectrum of contemporary information business at large. San Francisco/San Jose is the most striking example in this regard.

Technology per se cannot be viewed separately from society at large. 'Just as technology does not come into being outside of the social, so the social does not come into being outside of the technological' (Crang *et al.* 1999, p. 2). As we shall see, the dialectic of technology socialization and social *technologization* does not refer only to the social dissemination of technology as a product, but also to the emergence of specific social cultures as part of the innovation and production processes of information technology.

3.2   Technology and Flows

'Modern society is a society on the move' (Lash and Urry 1994, p. 252). Movements in both domestic and international space are conventionally classified into four basic types: goods and services, people, capital, and information, of which the first two are mostly concrete, whereas the two latter ones are normally invisible (see Kellerman 1993a). Traditionally, the geographical conception and measurement of these movements is by origins and destinations. At the domestic level, cities and regions are assumed to be engaged in such exchanges, which are reflected in their spatial organization, whereas international traffic is categorized through countries, regional blocs (e.g. the EU), or by global cores and peripheries (e.g. North America versus Africa) (see Smith 1993, p. 111). Intensified international movements of all four types have brought some analysts to suggest viewing the world as consisting of *region states*, rather than nation states, reflecting regional bonds within the global economy (Ohmae 1993).

All four movement types are longstanding, but their relative importance has changed as a result of developments in transportation and telecommunications media. Until the twentieth century, exports and imports dominated the international exchange arena. The development of aeroplanes increased the importance of international tourism, whereas the more recent emergence of direct and reasonably priced international telecommunications has assisted in the growing international traffic of information and capital.

Transportation and telecommunications are technologies of immense spatial and temporal significance, mediating between time and space, as well as between the local and the global. As an agent and facilitator of location, flow, and spatial change, technology may permit the controlling of resources, time, space, and production. As far as space is concerned, technology serves as the major facilitator for accessibility and for the overcoming of friction of distance (see Janelle and Hodge 2000). However, technology may not only control *experienced space*, but, via information technologies, *imagined space* as well. Hence, technology is used and expressed in the daily lives of both individuals and society at large.

The industrial revolution brought about a decline in agricultural employment, coupled with intensified agricultural production, so that human dependency on space as an active resource for agricultural production decreased. At the same time, innovations in transportation and communications technologies enabled a more extensive usage of space, and thus reduced human dependency on location, or on space as a passive or locational resource (Kellerman 1989). Through the introduction of railways, outlying areas were incorporated with core areas into national economic systems, but at the same time, the importance of terminal locations increased at the expense of travelled space (Schivelbusch 1978, see also Thrift 1996, pp. 264–267, 272–274).

The introduction of the telegraph in the mid-nineteenth century marked the separation between the transmission of physical objects and information. It further permitted the use of telecommunications to control transportation (Carey 1989, p. 203). The transmission of information through telecommunications no longer required a moving container, and has amounted, therefore, to an extensive distanciation of the spread and flow of information (see Giddens 1990). The telegraph further marked a change from a time-based society focusing on exchanging information relating to home places, to a space-based one, in which information on many places becomes available in a single time (Brooker-Gross 1985). Thus both social and commercial reach became more extensive (Carey 1989, p. 162, Marvin 1988, p. 202). Universal global telecommunications was predicted already in 1880 (Marvin 1988, p. 194). When it finally materialized in the late twentieth century it not only further decreased the importance of space as distance, but of the terminals involved as well, since the same information is now available simultaneously everywhere.

These innovations further served the expansionist nature of capitalism, permitting more extensive circulations of commodities and capital (Harvey 1985, p. 36). Increased demand for accessibility may yield transportation innovations, which, on their part, may contribute to time–space convergence, higher levels of locational utilities, and, hence, to a reorganization of space (Janelle 1969).

Contemporary technology plays a crucial role in making fusion processes among communications media feasible, and it may affect social attitudes to space through the three avenues proposed by Lefebvre (1991) and Harvey (1989), namely perception, experience, and imagination (Table 3.1). Technologies of transportation and telecommunications have permitted the geographically more extended productions and consumptions of both the spaces of places and the spaces of flows (see Kellerman 1989). The introduction of information technologies and particularly imaginative virtual spaces has led to the perception of spatial barriers as collapsing, notably at the global and national levels. Nineteenth-century transportation technologies were perceived as annihilating space *and* time (Schivelbusch 1978, Marvin

Table 3.1 Technology and social space

Human relation to space	Technological facilitation through transportation or telecommunications
Experience	Production of physical spaces and places
	Production of spaces of flows
	Space consumption
Perception	Lowering of spatial barriers
	Elevation in the importance of time versus space
Imagination	Creation and transmission of images of space and place

1988), whereas late twentieth-century telecommunications technologies led to the metaphor of the annihilation of space *through* time (Harvey 1989, see also Kirsch 1995). The very notions of time–space compression, coupled with globalization and the evolution of spaces of flows are not completely new trends (Thrift 1995). However, improved intranational and international transportation and telecommunications networks have facilitated more flows, especially instantaneous flows of capital and information, so that old spatial barriers consisting of international and interregional borders, as well as distance and time have, in many cases and instances, changed, shrunk, or even collapsed.

3.3 Knowledge ⇒ Innovation ⇒ Technology

The development and production of information technology is most heavily dependent on knowledge and innovation. As mentioned already in Chapter 1, the roles of knowledge and innovation, which are striking as far as information technology is concerned, extend far beyond information technology, and technology in general (see e.g. Bryson *et al.* 2000 pp. 1–2, Malecki 2000a). 'Knowledge and its applications for market success has become the primary source of competitiveness in the developed economies of the world' (Hotz-Hart 2000, p. 432). Therefore, 'in the competence perspective, the firm is essentially a repository of skill, experience and knowledge, rather than merely a set of responses to information or transaction costs' (Malecki 2000a, p. 107). Hence, the so-called *New Economy* was defined as: 'a global knowledge and idea-based economy where the keys to wealth and job creation are the extent to which ideas, innovations and technology are embedded in all sectors of the economy – services, manufacturing, and agriculture' (Atkinson and Gottlieb 2001, p. 3).

It was estimated for the U.S. that 80% of the value added in its manufacturing in the 1950s could be attributed to material products of all kinds and only 20% to knowledge, whereas by 1995 these proportions changed to 30% and 70% respectively (Stewart 1997). It is, however, misleading to assume that the role of capital has declined. As we shall see later, R&D activities are heavily capital-dependent, whether in the form of public research monies, commercial in-company capital, or mobilized venture capital. Capital is further required for the education and continual training of workers. Capital spending in the U.S. on information technology was, in 1965, just one-third of that of capital spending on product technology, exceeding it in 2000 (Dunning 2000b).

3.3.1 Knowledge for innovation

Knowledge is produced almost anywhere, and typical centers for knowledge production are universities. However, not every place of knowledge production constitutes by definition a place or center of innovation, notably a technological one. Various conditions regarding knowledge, capital and entrepreneurship are required for a place or firm to become innovative. Hence, in this section we shall focus on knowledge for innovation. The knowledge types and knowledge sources which serve as the basis on which technological innovation may emerge and flourish are varied. 'Not only is knowledge a heterogenous commodity and can be put to multiple uses; often, one kind of knowledge needs to be combined with several other kinds to produce a particular good or service' (Dunning 2000b, p. 9). The various types of knowledge required for innovation processes are complementary and indispensable (Antonelli 2000a,b). Seven types of these are outlined below:

(1) In-house tacit knowledge: Tacit knowledge is by far the most important type of knowledge required for innovation to take place, with high importance attached to it all the way through, from basic R&D to early production stages (Audretsch 2000, p. 72). In-house tacit knowledge becomes available through well-educated and well-trained human resources within an innovating firm, or within a university or research organization. It is, therefore, firmly embedded within organizations. However, the transformation of academic knowledge into a viable commercial one within universities has its own difficulties (Luger and Goldstein 1991).

(2) In-house 'innovation star(s)': In her discussion of the biotech industry, Feldman (2000 p. 381) defined a 'star scientist' as 'a highly productive individual who discovered a major breakthrough'. Innovation leaders are actually required within almost any R&D team or within an innovative company, even if many of these leaders are at a more modest scale of 'starring'. These stars not only embody knowledge and ideas, but they lead others in innovative thinking and scientific–technological visioning and daring.

(3) Codified knowledge: Knowledge available in texts of whatever form, whether freely available or for a price, are an important knowledge resource for innovation, even if those involved in the innovation process have already completed their formal studies. Major university libraries or research institutions are of high significance in this regard.

(4) In-region knowledge spillover: This spillover consists of formal and mostly informal transmission of knowledge, notably a tacit one, among people engaged in innovation processes, as well as similar transmissions among firms. A region, in the sense of the geographical framework for knowledge spillover, may differ widely in scale, ranging from a single science park to a whole metropolitan region, depending on the size and spread of the local high-tech industry. Knowledge spillover may be considered second to in-house tacit knowledge in its importance for the innovation process, serving as *knowledge externalities* (see e.g. Audretsch and Feldman 1996). On-going, daily transmission of knowledge takes place within regions and the special significance of regions in this regard will be dealt with in the next section.

Like information, the transmission of knowledge from one person to another does not eliminate the transmitted knowledge from the transmitter (Malecki 2000a). Innovative firms have a conflict of interest in this regard. On the one hand, they would not like to have knowledge gained within their companies spill over to other companies. On the other hand, they appreciate the tremendous value, for their interests, of knowledge transmitted into their firms, as well as brainstorming through knowledge exchange. The codification of knowledge cannot substitute for face-to-face exchange of tacit knowledge. Even institutional attempts, such as the recent MIT project to post, on a free Internet site, all the course materials of their degree programs, only emphasize the extreme importance of class and lab. experiences for degree certification.

(5) Extra-regional knowledge spillover: Knowledge spillover beyond the geographical limits of a given region, whether metropolitan or another, is of a different and more limited nature, but it may carry benefits and significance, even beyond the innovation process per se. Extra-regional spillover processes may take several forms. First, *connections* with colleagues in other areas or the emergence of *innovation networks* (Camagni 1991), which may be maintained through professional meetings or through the Internet (see Malecki 2000a). Second, *trade*, notably international trade, may bring with it knowledge spillover, coupled with enhanced competitiveness and outputs (Storper 2000a, Ben-David and Loewy 1998). Third, *globalization* of innovation activities. This could take the form either of globally spread R&D activities of multinational corporations (MNCs), or through knowledge export. Hotz-Hart (2000) counts among the advantages of global R&D the access to globally spread but locally pooled knowledge. Global R&D also may carry a loss of a critical domestic-regional R&D mass and synergy, coupled

with high coordination costs for the global R&D activities. Thus, as Patel and Pavitt (1991) showed, the technological activities of large firms in most developed countries in the 1980s was rather limited.

(6) Continuous learning: An innovative firm or organization must be involved in continuous learning and adjustment processes in order to thrive through the creation of new ideas, as well as improving existing ideas and products. These imply continuous learning by employees, as well as continuous hiring of recent university graduates in relevant fields. It further implies a firm policy of openness to required organizational changes. In all these aspects of continuous learning the regional setting is of importance, as we shall see in the next section.

(7) Business knowledge: This type of knowledge includes producer services which can be bought by innovative firms through outsourcing, or they may be located, partially at least, in-house. Among the most important services, beyond capital mobilization, are marketing and market research experts, test laboratories, patent attorneys and other legal services related to innovation and R&D activities (see Feldman and Florida 1994). Though lying outside the knowledge/innovation formation, cumulation, and exchange arena, access to tacit knowledge in these fields is of critical importance for an innovation to mature into a marketable product.

Business knowledge implies not just the way a business should be conducted and marketed. It implies also entrepreneurship, which is most vital for innovativeness at large, and for technological innovation in particular. 'Entrepreneurialism, as an essential dimension of the Internet culture, comes in with a new twist: it creates money out of ideas, and merchandise out of money, making both dependent on the power of the mind' (Castells 2001, pp. 59–60). Entrepreneurship is not just something that has to be acquired through learning. It requires also a drive for risky innovation, as well as a general entrepreneurial atmosphere. The role of government is crucial in this regard, through the creation and encouragement of proper taxing systems, fostering of financial tools such as venture capital funds, creation of R&D funds, and the promotion of a social appreciation of risk-taking entrepreneurs.

3.3.2   Regions of knowledge and innovation

The seven types of knowledge for innovation have to be localized within a geographical setting in order to yield innovativeness. This setting is the region, normally a metropolitan area, or a metropolitan area with additional areas beyond it. For some commentators, the region, or geography at large, constitutes a mere *vessel* for the innovation process (Feldman and Florida 1994, p. 210), whereas others see it more actively: 'the *territory* itself plays a role, as a place of co-ordination and learning' (Malecki 2000a, p. 113) A recent analysis of European firms has shown the importance of the firm itself, which may supersede that of the region (Sternberg and Arndt 2001). The region ties the required knowledge together, and facilitates in-region spillover within a localized embeddedness of an idiographic nature. This is achieved through two major elements: social capital and industrial organization.

'Social capital refers to the values and beliefs that citizens share in their everyday dealings and which give meaning and provide design for all sorts of rules' (Maskell 2000). These rules include norms, codes, trust and solidarity, among other things. The untradeable asset of social capital serves as a basis for companies sharing the same region when it comes to exchanges of tacit knowledge. Social capital may obviously develop and exist in regions which are not necessarily specialized in IT R&D and production. However, social capital, or region-specific trust and codes are crucial for knowledge exchange and sharing.

A favored regional industrial organization for the development of knowledge spillover is one of the three types of the so-called *sticky places*, or attractive regions, namely the *Marshallian industrial district* (Markusen 1996). This type of industrial district consists mainly of locally-owned small firms. In such a region, workers feel more committed to the region than to their firm, and thus lean more to tacit knowledge exchange.

Sustaining the advantages of knowledge spillover is a continuous regional task, through continuous knowledge gaining, and hence the term *learning region* (Florida 1995, Jin and Stough 1998, Maskell and Malmberg 1999b). However, if the level of innovation is measured through the number of patents achieved per city, then it was shown for the U.S. that city size and historical industrialization determine the number of patents (Ó hUalacháin 1999).

Knowledge exchange may be achieved in various ways. First, and most important, are face-to-face meetings, and repeated ones, in order for knowledge to be disseminated (Audretsch 2000). Firms may engage in various forms of exchange, beginning with simple *barter*, and moving through *dyadic*, semi-stable exchange, to *network-relations* (Maskell 2000). These may be enhanced via electronic communications, as well as through strong contacts with local universities (Antonelli 2000a). Another mechanism for knowledge spillover is patent citations, which tend to be more extensive between firms and universities located within the same region (Jaffe *et al.* 1993).

These mechanisms of tacit knowledge spillover bring about the clustering of high-tech innovative firms, so that the more innovation-oriented a firm is, the more it tends to locate in the proximity of similar firms, as the spillover effect is restricted in distance (Dunning 2000b, Audretsch and Feldman 1996;). Knowledge transfer may become stronger when firms of various specialties, based on similar scientific knowledge, cluster in the same region (Feldman and Audretsch, 1999). By the same token, small firms may enjoy knowledge spillover from nearby large firms (Feldman, 2000). It is questionable whether globalization increases regional clustering and regional concentration or just the opposite (Sölvell and Birkinshaw 2000, Dunning, 2000b).

3.3.3   The innovation process

The innovation process is described in Figure 3.1, and though it may apply to innovation at large, it is of special relevance and significance to IT. Mansfield (1991) estimated some ten years ago that the time lag between an academic research finding and production based on it was seven years, so that the innovation process described in Figure 3.1 would represent a 10–12 year span. It is reasonable, however, to assume that the intensified R&D processes in the 1990s have shortened these time spans.

Innovation may be motivated through two different motives leading to R&D, which are normally located within two separate types of organization. Market motivation typifies commercial companies, small or big, whereas an intellectual one may motivate universities and research centers. This difference is blurring, however, with universities being pushed towards commercialization of their research knowledge. As noted already, the geographical proximity of the two within a regional setting may benefit both. The innovation process involves several investments and costs, but the initial one, the financing of R&D, usually constitutes the heaviest and riskiest. Sources for financing depend on the organization leading the innovation process. Universities may receive most of their research funds from their own and public foundations, but they also may receive research funds from the industry, although many universities still refrain from doing so. The commercialization of university research brings about negotiated partnerships, on a project by project basis, between them and business organizations, notably at later stages of the innovation process.

Private industry may finance R&D from corporate funds, or through venture capital, mobilized directly or through dedicated funds. During 1995–2000,

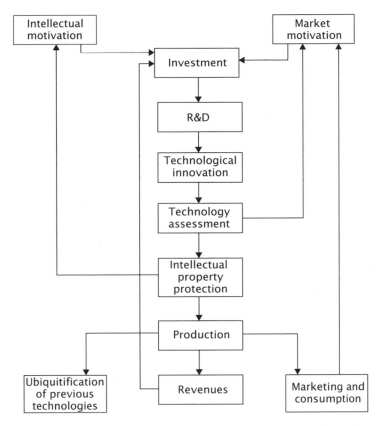

Figure 3.1 The innovation process: The technology perspective (see also Mitchelson 1999, Figure 4)

the peak years of venture capital investments in the U.S., the total investments increased by 1300%, and investments in the Internet amounted to 90% of the growth 1998–2000 (Zook 2002). In some countries government may assist directly innovative industrial R&D, on the assumption that successful production will pay back royalties. Such a process assists in the generation of employment, as well as continuous knowledge development and spillover.

Successful R&D, as part of an innovation process, implies reaching a technological innovation. Such an innovation, whether achieved through academic or commercial research, requires an assessment of its development into a product vis a vis the market, in order to reach a production decision.

Such an assessment of an innovation may lead to either further development of the innovation in order for it to fit market needs, or it may be suspended. Once finalized, the innovation has to be protected, either through patenting, or through licensing in the case of software which is not patented. Patenting may serve as a starting point for additional innovative research, mostly within universities. Universities may sell patents developed by their faculty, following a technology assessment by potential buyers.

The phases described so far normally take place within one organization, whether academic or commercial, located in one region. Production may potentially take place elsewhere. However, since technological products involve complex processes before they become standardized, production at early stages takes place within the R&D region or close by, thus enhancing the development of these regions, and further permitting knowledge spillovers from the production process as well (Audretsch and Feldman 1996). Once standardized, production may move to other areas, either fully or partially, notably to regions specializing in mass production by large companies (Felsenstein 1993, Markusen 1996).

Production may yield three results. First, marketing and consumption, which may bring about a new round of market motivation for innovation, if the use of the new technology brings about new demands or potential demands for new products. Second, revenues from successful marketing of the new product may finance new rounds of R&D. Zook (2002) found that in the Internet industry, venture capital is invested in existing knowledge clusters and in a cyclical process it leads in turn to further innovations. Finally, new technologies may bring about a *ubiquitification* of older knowledge, technologies and products (Maskell and Malmberg 1999a). For a company to keep its competitiveness, it has, therefore, to be continuously involved in an innovation process.

3.3.4 The innovation process and the regional economy

Information technologies were defined as 'a specific form of productive organization, deriving their specificity from the distinctiveness of their raw material (information), and from the singularity of their product (process oriented devices with applications across the entire spectrum of human activity)' (Castells 1989 p. 71). Local concentrations of high-tech industries, in the form of planned industrial parks, were termed *technopoles* (Castells and Hall 1994), whereas a regional concentration of such industries was called *technopolis* (Scott 1993). Sometimes, emphasis by governments is on the development of proper telecommunications and IT infrastructure, creating *intelligent corridors*, notably in Singapore and Malaysia (see Corey 2000). From the late 1990s the national scale has become increasingly significant for the development of high-tech industry, as numerous nations, notably small ones, have developed national policies towards high-tech R&D and/or production (see e.g. Lundvall and Maskell 2000). Sweden and Finland are more veteran nations in this regard, whereas Ireland, Singapore and Israel constitute more recent examples. The leading *technopoli* worldwide will be reviewed in the next section.

The emergence of *technopoles* (and of *technopoli* at a later stage) is based on three macro-processes: the rapid developments in information technology; the emergence of a global economy; and the evolution of the informational economy, based on the generation of new knowledge. These three processes are interconnected (Castells and Hall 1994, pp. 3–4). The sweeping growth of the Internet has brought about new rising industries during the 1990s, notably in the form of start-up enterprises, coupled with changes in the production scheme of existing industries, such as the telecommunications industry. The early 2000s signify a similar process regarding mobile tele-phony, despite the crisis in the NASDAQ stock market, serving as a major vehicle for venture capital mobilization for the industry.

From the regional/industry perspective, it is possible to trace the following process-chain for the high-tech industry, leading from knowledge and capital formation, through innovation and invention, to production (Figure 3.2) (see Feldman 1994 pp. 1–2). The industry is based on three major production fac-tors: scientific and technological knowledge, market knowledge, and venture capital. As shown in previous sections, abundant scientific and technological knowledge, in the form of well-trained human resources, coupled with the proximity of major research universities and research institutions, is the most important locational factor at all levels: local, regional and national. Scientific-technological human resources are still the relatively least loca-tionally flexible production factor, in terms of immigration, notably at the international level. This inflexibility is despite the globalization of profes-sional human resources and the attraction of major global R&D centers, such as Silicon Valley (see Castells 2000 p. 130). The multitude of workers of different national origins in such centers gives a misleading view regarding the cultural-national attachment of the majority of trained workers who stick to their home countries.

As we have seen already, market knowledge refers both to the under-standing and assessment of demand for new ideas and products, as well as to the entrepreneurship required for the development of new products and

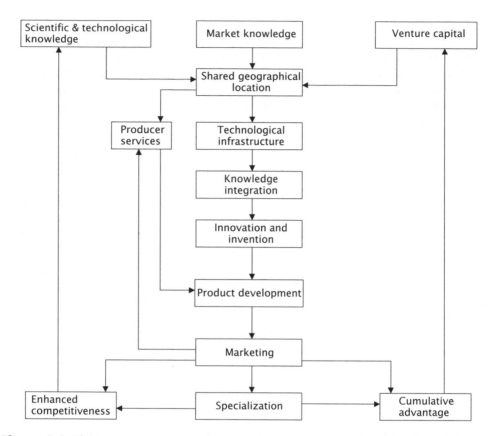

Figure 3.2 The innovation process: The regional perspective (see also Crevoisier and Maillat 1991, p. 15; Sternberg and Arndt 2001, Figure 1)

their marketing. This knowledge does not necessarily have to be locally or domestically based for each enterprise, though profound levels of market knowledge serve as prerequisites for the emergence of national or regional loci of innovation.

Venture capital has been assumed by various writers to be a locationally inflexible factor, so that it has to be regionally embedded and mobilized (DeVol 1999, Cooke *et al.* 1998). In the mid-1990s, both the U.S. geographical sources for venture capital and their investment were highly concentrated (Smith and Florida 2000). Thus, the three metropolitan areas of New York, San Francisco, and Boston accounted for 50% of all sources; the three corresponding states for 64%; and the two corresponding regions (the Northeast and the Pacific) for 75% of all venture capital resources. The investments were concentrated in the Silicon Valley which received over 20% of national investments (amounting to 50% of the investments in California, with an additional investment region in Los Angeles), and in Route 128 in suburban Boston which received close to 9% of national investments (amounting to 95% of investments in Massachusetts). New York and Chicago were thus exporters of venture capital, but the two major receiving areas, Silicon Valley and Boston raised resources within their own regions as well. Another study found the share of Silicon Valley in the number of venture capital investments in the U.S. in between two later years, 1997 and 1998, to amount to 27.2%, followed by Boston (11.8%), New York (6.8%), and Los Angeles (4.4%) (Zook 2002). New York has thus become a center of investment as well. Venture capital was generally found to be tending to go to areas with previously well-established high-tech industries.

The increased globalization of capital flows has permitted investments of venture capital at a global scale, notably as of the mid-1990s, so that of the three major production factors this has become the geographically more flexible one (see Kellerman 2002a). Venture capital is of extreme importance to the Internet industry and its competitive first mover accent, since this type of capital moves with much speed and since it relies on the wider regional environment of high-tech business (Zook 2002).

For a *technopole* or *technopolis* to take off, local availability of producer services of all kinds, notably telecommunications and finances, is crucial. Producer services play a major role for both the domestic and global activity of the industry, notably at the market development phase. This is also the sector that may benefit most from the innovation to production process chain, as it serves both the innovation and production phases (DeVol 1999).

Knowledge integration, facilitated through the availability of capital and producer services, may finally mature into innovations and inventions. This integration was termed by Castells (1989), the *milieu*, namely 'a specific set of social relationships of production and management, based upon some common instrumental goals, generally sharing a work culture, and generating a high level of organizational synergy' (p. 72), and 'an industrial complex becomes a *milieu* of innovation when it is able to generate within itself a continuous flow of the key elements that constitute the basis for innovative production of information technologies, namely new scientific and technological information, high-risk capital, and innovative technical labour' (p. 88) (see also Camagni 1991, Cooke *et al.* 1998, Garnsey 1998, Antonelli 2000b).

67

As we have noted already, the industrial production of products based on new innovations does not necessarily have to take place in the location or country of innovation, though initial experimental production is normally adjacent to R&D location. From a regional perspective, if innovation, production and marketing all share the same location, they may bring about enhanced competitiveness, specialization, and cumulative advantage, which on their part may enhance the conditions for new rounds of innovation (see Pred 1977, p. 90, DeVol 1999, Shefer and Frenkel 1998). Moreover, this specialization may attract additional innovative firms (Malecki 2000a). There is only partial evidence on possible geographical decentralization of production, even within metropolitan areas (Malecki 2000a, Hackler 2000).

Through continuous learning and resulting innovation rounds, high-tech regions constantly engage in the enhancement of their *sustainable advantage*, through continuous betterments of technology and human resources (Florida 1995). The innovation process and landscape described so far in this and previous sections, may be concluded by Feldman and Florida's (1994, p. 226) assessment that it constitutes 'agglomerated and synergistic social and economic institutions welded into a technological infrastructure for innovation'.

3.4   Information Technology Regions

The discussion of knowledge and innovation so far has been rather theoretical and conceptual. In this section, we shall examine information technology centers worldwide. The first discussion will present a global distribution of such centers, whereas the second one will focus on some specific countries and regions.

3.4.1   IT regions worldwide

Camagni (1991, pp. 4–5) and Castells (2000 p. 422) contended that *milieux of innovation* worldwide are interconnected, despite their spatial discontinuity, and that they cooperate with each other within networks of interaction, despite the obvious competition among them. Such a cooperation exists mainly in the exchange of knowledge, as well as in exchanges of information on market and capital mobilization. These may turn IT regions into a global network. IT regions worldwide are presented in Tables 3.2–3.4, and in Figures 3.3 and 3.4. The data set used for these tables and figures, entitled *Venture Capitals*, was compiled by Jennifer Hillner (2000) for *Wired*, just

Table 3.2 Leading high-tech centers

Center	Universities & research	Established companies	Entre-preneurship	Venture capital	Total score
Albuquerque, NM, U.S.	4	3	3	2	12
Austin, TX, U.S.	3	4	4	2	13
Baden-Würtemberg, Germany	3	3	2	2	10
Bangalore, India	3	4	3	3	13
Bavaria, Germany	3	3	2	3	11
Boston, MA, U.S.	4	4	3	4	15
Cambridge, U.K.	4	3	3	2	12
Campinas, Brazil	4	3	1	0	8
Chicago, IL, U.S.	3	2	2	2	9
Dublin, Ireland	3	3	3	3	12
El Ghazala, Tunisia	1	1	1	1	4
Flanders, Belgium	4	2	3	2	11
Gauteng, South Africa	1	1	1	1	4
Glasgow-Edinburgh, U.K.	3	3	1	1	8
Helsinki, Finland	3	4	4	3	14
Hong Kong, China	3	2	2	2	9
Hsinchu, Taiwan	3	1	4	3	11
Inchon, South Korea	3	2	2	2	9
Israel	4	4	4	3	15
Kuala Lumpur, Malaysia	2	3	1	2	8
Kyoto, Japan	4	1	3	3	11
London, U.K.	4	3	3	4	14
Los Angeles, CA, U.S.	3	3	2	3	11
Malmö, Sweden-Copenhagen, Denmark	3	3	2	3	11
Melbourne, Australia	3	2	3	2	10
Montreal, Canada	3	4	2	3	12
New York City, NY, U.S.	3	3	3	3	12
Oulu, Finland	3	2	3	2	10
Paris, France	n.a	n.a.	n.a.	n.a.	10
Queensland, Australia	2	3	2	2	9
Raleigh-Durham-Chapel Hill, NC, U.S.	4	4	3	3	14
Salt Lake City, UT, U.S.	3	2	2	1	8
San Francisco, CA, U.S.	3	3	3	4	13
Santa Fe, NM, U.S.	3	2	2	1	8
Saõ Paulo, Brazil	1	3	3	2	9
Saxony, Germany	3	2	1	2	8
Seattle, WA, U.S.	3	4	3	2	12
Silicon Valley, CA, U.S.	4	4	4	4	16
Singapore	1	2	2	2	7
Sophia Antipolis, France	2	3	2	1	8
Stockholm-Kista, Sweden	3	4	4	4	15
Taipei, Taiwan	4	3	3	3	13
Thames Valley, U.K.	3	3	2	2	10
Tokyo, Japan	3	2	3	3	11
Trondheim, Norway	2	1	2	1	6
Virginia, U.S.	3	3	2	2	10

Source: Hillner (2000).

Table 3.3 High-tech centers with leading scores

Universities & research	Established companies	Entre-preneurship	Venture capital	Total score
Albuquerque,	Austin	Austin	Boston	**16**: Silicon Valley
Boston	Bangalore	Helsinki	London	**15**: Boston
Cambridge	Boston	Hsinchu	San Francisco	Israel
Campinas	Helsinki	Israel	Silicon Valley	Stockholm
Flanders	Israel	Silicon Valley	Stockholm	**14**: Helsinki
Israel	Montreal	Stockholm		London
Kyoto	Raleigh-Durham			Raleigh-Durham
London	Seattle			**13**: Austin
Raleigh-Durham	Silicon Valley			Bangalore
Silicon Valley	Stockholm			San-Francisco
Taipei				Taipei

Source: Hillner (2000).

Table 3.4 High-tech center distribution by continent and country

Continent and country	No. of centers	Score range
North America	14	8–16
Canada	1	
U.S.	13	
Europe	16	6–15
Belgium	1	
Finland	2	
France	2	
Germany	3	
Ireland	1	
Norway	1	
Sweden & Denmark	2	
U.K.	4	
Pacific	11	7–13
Australia	2	
China	1	
India	1	
Japan	2	
Malaysia	1	
Singapore	1	
South Korea	1	
Taiwan	2	
Central and South America	2	8–9
Brazil	2	
Africa	2	4
South Africa	1	
Tunisia	1	
Middle East	1	15
Israel	1	

Source: Hillner (2000).

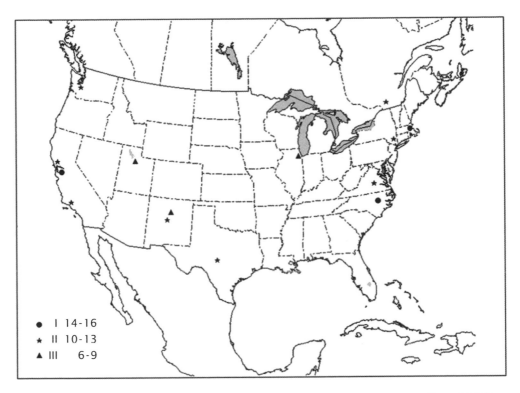

Figure 3.3 High-tech centers in North America, 2000. (Data source: Hillner 2000)

before the NASDAQ crisis. It presents 46 high-tech centers comprising mainly IT R&D and production, as well as biotechnology. The centers were ranked by Hillner through a 1–4 scoring for each of the following four factors: 'the ability of area universities and research facilities to train skilled workers or develop new technologies; the presence of established companies and multinationals to provide expertise and economic stability; the population's entrepreneurial drive to start new ventures; and the availability of venture capital to ensure that the ideas make it to market' (Hillner 2000, p. 1).

The use of subjective scores ranging only between one and four, as well as the omission of any quantitative size variables, such as the total number of workers, the number of scientific and technical workers, the number and sales of firms, etc., is problematic. Thus, for example, Chicago, which is leading in the total number of high-tech workers, received a relatively low

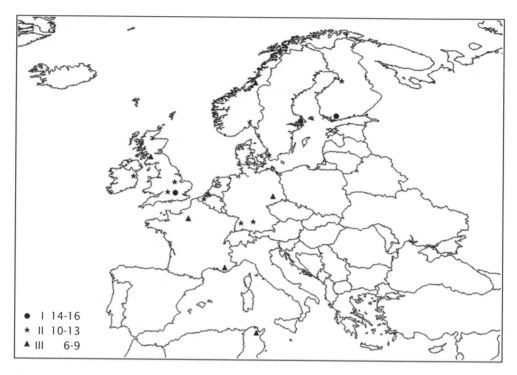

● I 14-16
★ II 10-13
▲ III 6-9

Figure 3.4 High-tech centers in Europe, 2000. (Data source: Hillner 2000)

total score. A comparison between Hillner's U.S. IT regions and their scores, and DeVol's (1999) list of *tech-poles* for the U.S., based on quantitative data (Table 3.5 below) reveals that eight out of the ten American leading tech-poles appear in Hillner's data as well, with high-ranking Dallas, and more modestly-ranking Atlanta missing. Also, San Jose scored extremely high in DeVol's (1999) list compared to following tech-poles, whereas the gap between Silicon Valley and following venture capitals in Hillner's (2000) list is more moderate. On the other hand, highly ranked Austin, Raleigh-Durham-Chapel Hill, and San Francisco, on Hillner's (2000) list, were ranked low (21, 18, and 22, respectively) on DeVol's (1999) list, so that both lists may be viewed as presenting rather rough and descriptive estimates.

As Table 3.3 shows, Silicon Valley leads in all four parameters, and actually all other centers were judged against it. Following are Boston (missing one point in entrepreneurship); Israel (missing one point in venture capital); and Stockholm (missing one point in universities and research). With the exclusion

Table 3.5 U.S. city rankings by technology indices

Patents (1996)	Innovations (1982)		Tech-poles (1998)		New economy (1998/9)	
New York	New York	735	San Jose	23.69	San Francisco	95.6
San Francisco	San Francisco	477	Dallas	7.06	Austin	77.9
Los Angeles	Boston	345	Los Angeles	6.91	Seattle	68.0
Boston	Los Angeles	333	Boston	6.31	Raleigh-Durham	61.4
Chicago	Philadelphia	205	Seattle	5.18	San Diego	61.4
Detroit	Chicago	203	Washington	5.08	Washington	60.6
Philadelphia	Dallas	88	Albuquerque	4.98	Denver	58.1
Minneapolis	Cleveland	77	Chicago	3.75	Boston	54.0
Dallas	Detroit	68	New York	3.67	Salt Lake City	49.8
Washington/	Pittsburgh	42	Atlanta	3.46	Minneapolis	49.0
Baltimore						

Sources: Patents: Ó hUallacháin (1999); innovations: Feldman and Audretsch (1999); tech-poles: DeVol (1999); new economy: Atkinson and Gottlieb (2001).

of Israel, it is for North American and European centers to lead the global system of high-tech centers, and only in the fourth level (of centers with a total score of at least 13 points), are there two Pacific centers, Bangalore and Taipei. Cumulative advantage in industrial experience, research, capital, and entrepreneurship gained by North American and European centers therefore plays an important role.

As for leadership in the specific parameters, it is interesting to note that the number of centers which ranked highest, with four points per parameter, is not equal for all parameters: universities and research (11); established companies (10); entrepreneurship (6); and venture capital (5). This may signify the relatively wide geographical distribution of knowledge, extending from rather veteran Boston and Cambridge to new Campinas and Taipei. The distribution of established companies is also relatively wide. However, the rather 'newer' elements of entrepreneurship and venture capital are still quite concentrated. This is particularly significant as far as venture capital is concerned, as it is located and invested highly in world cities which serve also as financial centers. This despite the potential global mobility of capital.

Table 3.4 demonstrates strongly the digital divide (to be discussed in Chapter 7). Europe is leading in the number of centers, whereas the North American ones present a higher scoring range. The three cores, North America, Europe, and the Pacific include 41 out of the 46 centers, with just two additional ones in Brazil, two very low ranked ones in Africa (in South Africa and Tunisia), and a high-ranking one in the Middle East (Israel).

Table 3.4 and Figures 3.3 and 3.4 present uneven geographical distributions of high-tech centers within North America and Europe. In North

America veteran industrial areas along the Atlantic and Pacific coasts fare better, though interior States such as New Mexico, Utah, and Texas are present. This type of pattern was noted also by Feldman and Audretsch (1999), as well as by Ó hUallacháin (1999), in their analyses of the geographical patterns of patenting and innovations. In Europe, northern countries dominate the scene, with only one Mediterranean center, Sophia Antipolis, which enjoys strong support by the French government. This reflects not only different levels of economic development, but also different policies, notably by smaller nations. Only one center, Saxony (Germany), is located in a previous East European territory. Thus, Gillespie *et al.* (2001, p. 119) noted the rather 'remarkable stability of the geography of innovation within Europe' based on veteran concentration of knowledge and capital. Within the Pacific, however, current intensive efforts for the development of high-tech R&D and production in China, including Hong Kong, may challenge the Japanese leadership in the future.

3.4.2   IT R&D and production in selected countries

National systems of production, knowledge and learning may bring about different patterns of high-tech innovation and production (see e.g. Lundvall and Maskell 2000). As we shall see, even within one large country, different patterns may evolve, as is the case for Boston and Silicon Valley in the U.S. (Saxenian 1994). These different patterns have brought about differences not only in levels of IT R&D and production, but in information production as well. Nations differ in their time perspective on inter-firm relations (relatively short in the Anglo-Saxon countries and long in Japan); organization of capital markets; interactions between industry and universities; education and training systems; norms, habits and rules welding into trust; as well as in the role of authorities (Lundvall and Maskell 2000).

Two newly evolving IT production areas (Bangalore, India, and the MultiMedia Super Corridor, Malaysia) were recently briefly presented by Graham and Marvin (2001, pp. 337–342). Our following exposition will highlight the U.S., the Scandinavian nations, and Israel. The first one being a large nation in both territory and population with the longest and most developed capitalist system, accentuating market domination and openness for innovations; the second being a multi-nation region with small and rather scattered populations applying special significance to the role of government in economic egalitarian development; and the third being a small nation in both territory and population, enjoying a unique accumulation of tacit knowledge.

U.S.: The U.S. may be considered not only the *founding nation* of information technology at the levels of development, production and dissemination,
but also the country with the largest number of IT centers. The U.S. is the
country studied by far the most extensively, regarding its own centers, as
well as innovativeness at large. At least four detailed surveys of American metropolitan economies vis a vis high-tech production have appeared
recently (Atkinson and Gottlieb 2001, Markusen *et al.* 2001; Cortright and
Mayer 2001, DeVol 1999). The U.S. has counted on a long tradition of innovativeness and entrepreneurship, as well as mature capital markets, all as part
of its capitalist system. It could further count on its extensive high-learning
and research system.

Ranking U.S. cities by technology measures depends on the measured
variable, and on the geographical definition of city boundaries. Table 3.5
presents four such measures in various spheres. Strictly defined innovations,
by protected registered patents, were reported by Ó hUallacháin (1999) for
1996, albeit without the number of patents per city. Feldman and Audretsch
(1999) used a softer measure of innovation, though using 1982 data. Their
measure presents new product introductions, as a direct measure of innovative output, rather than using patents, which they consider intermediate
outputs. Moving to the production side, DeVol (1999, p. 65) defined *tech-
poles* as cities serving as technology production centers, measured by an
index consisting of location quotients and center shares of national high-
tech output. The *new economy index* proposed by Atkinson and Gottlieb
(2001) goes one step further, incorporating into the index variables relating
to high-tech production and consumption, as well as to the overall economy.
Thus, their 16 variables related to five areas: knowledge jobs; globalization;
economic dynamism and competition; the transformation to a digital economy; and technological innovation capacity. Their general results show that
cities devoted to high-tech rank high, regardless of their size (Los Angeles
was ranked 20, and New York 17).

The data in Table 3.5 reveal the unique role of the San Francisco/San Jose,
or the Silicon Valley area, in the U.S. high-tech economy. Ranking fifth in
population in 1996, and producing less than 6% of the high-tech output
(DeVol 1999), the area has ranked second in innovation, first in production,
and first in an overall measure of the information economy. The gaps between
the San Francisco rates and the following cities are significantly wide. On the
one hand, this finding attests to the uncontested leadership of the Silicon
Valley area, giving the impression that information technology innovation
and production in the U.S. is still concentrated. On the other hand, however,

75

many other cities have become active and even specialized (e.g. Austin) in high-tech innovation and/or production.

Two additional cities deserve some attention. New York turns out as a leader in innovation rather than in production. It should be noted in this regard that patents are registered many times in the city where company headquarters are located (Ó hUallacháin 1999). On the other hand, New York is a true leader in innovation, not only in technology, but also in technology business and organizational applications. Boston was ranked number 3–4 as an innovative and producing city, but given the world-known high-tech concentration along Route 128, its ranking might have been higher. Saxenian (1994) and Malecki (2000a) noted the unique culture of the Boston area, typified by a high concentration of universities, especially when compared to Silicon Valley. Whereas Silicon Valley has become typified by its becoming a networked and rather flexible knowledge-sharing industrial area, thus encouraging new entrepreneurs, the Boston area has emerged as an industrial area consisting of larger, rather 'closed' companies. Smaller companies, established in the region, imitated the work style of the larger ones. The rather open nature of the Silicon Valley has assisted not only its flourishing, but also its coping with crises. In the Boston area, however, it is for the heavy influx of knowledge from universities in the region to provide for both the ability to flourish and to cope with crises.

Using wider employment variables for a much wider definition of industries with high innovative content, including architectural services and management and public relations, Markusen et al. (2001) were able to reach a completely different ranking from data on the overall job gains in the 1990s. Chicago headed the list, followed by Washington, DC, Silicon Valley, Boston, New York, Philadelphia, Dallas, Seattle, Minneapolis/St. Paul, and Houston, in this order.

Scandinavia: Maskell (1999) points to several Nordic traditions which might explain the early transition of Scandinavian nations into mature capitalist national economies, despite their small size and earlier specialization in staple and wood exports. Egalitarian income distribution across social strata coupled with an aversion to conspicuous consumption have permitted capital direction to productive uses. Nordic societies have further been typified by consensus seeking in a society with shared cultural heritage and sense of unity. In addition, countries in the region developed a strong tradition for public intervention in economic development. The Finnish labor force was described as putting a special emphasis on continued education (Brunn and Leinbach 2000).

These traits have been significant in the development of the telecommunications industry in Scandinavian countries (Kellerman 1999a). The Swedish major telecommunications company *Ericsson* was founded in the 1870s and had already begun to manufacture telephones in the 1880s. In Finland, the first telephone was installed in 1877, one year after its invention by Alexander Graham Bell in the U.S. *Ericsson* began exporting before World War I, and realized that in order to keep export prices low, the production price per unit should be kept low as well. This was achieved by low profit margins in the domestic market which resulted in expanded demand (Ahnstrom 1997). The Finnish *Nokia* company has developed similarly (Roos 1994), though it moved into the telecommunications business mainly during the last thirty years, aided by government, and has turned into an influential *regional fountainhead* (Maskell *et al.* 1998, Brunn and Leinbach 2000).

State monopolistic service providers cross-subsidized tariffs in the peripheries by higher ones in urban areas. This infrastructure and the high profits of the state monopolies assisted later on in the rapid and reasonably-priced evolution and penetration of cellular telephony and in the creation of equal standards in all Nordic countries, so that the same device could be used everywhere in the region. The two companies, *Ericsson* and *Nokia*, were able to lead the cellular telephone industry also because of more flexible regulation policies in Nordic countries, emerging earlier than in other countries (Maskell *et al.* 1998).

These same attitudes to the role of government have assisted the development of high-tech centers in remote areas, such as Trondheim (Norway), and Oulu (Finland), in addition to major metropolitan areas, such as Stockholm, Copenhagen and Helsinki, which serve also as major financial centers and gateways, as well as concentrations of producer services. The Nordic high-tech industry has further enjoyed 'shared trust and exchange of tacit knowledge across firms as well as between business and government on a mainly informal level' (Maskell *et al.* 1998, p. 153).

Israel: The Israeli case is worth attention, not only because of the significant size of the knowledge, innovation, and production components of its high-tech sector, but because of its location, outside the three global economic cores (Kellerman 2002a). The deeper roots for the Israeli high-tech industry lie in Jewish cultural traits of scholastic learning, as well as in prolonged traditions of international communications among Jewish communities in the Diaspora, rather than in a long tradition of industrial development. Within a modern context, the industry based itself through a strong security accent attributed to the Arab–Israeli conflict.

Table 3.6 Phases in the development of the Israeli high-tech industry

Phase	Decade	Characteristics	R&D accent
I	1960s	Early domestic beginnings	Security
II	1970s	Investments by American multi-nationals	Semiconductor R&D in large plants
III	1980s	Crisis and restructuring	First Israeli civilian innovations (e.g. printing, CT)
IV	1990s	Domestic technological entrepreneurship and global capital investments	R&D and innovative production by start-up companies

Source: Kellerman (2002a).

Since the mid-1970s the Israeli economic system has undergone a gradual transformation from a mixed socialist–capitalist system into a market system, a transformation led by all Israeli governments since then (see Kellerman 1993b). Competitiveness has been promoted, as well as less governmental involvement and stronger commercial and financial international ties. These trends have encouraged high-tech entrepreneurship and investments in the production of information technology, as well as in services for information provision. It is possible to identify four phases in the emergence of the Israeli high-tech development (Table 3.6). These phases moved from the creation of a domestic base for R&D in information technology to an increasingly global orientation, as well as from an emphasis on large firms, domestic as well as American ones, to small start-up companies in the 1990s. These reached 3000 in number in 2000, compared to 10 000 in the U.S. in the same year. The Israeli high-tech industry presents a very high domestic accumulation of knowledge, embodied in academic research, as well as in the world's largest per population rates of scientists (2.66 per 1000 population in 1987), engineers and technicians (140 per 10 000 in the mid-1990s). This accumulation of a well-trained workforce reflects army training, notably in intelligence services, as well as migration to Israel of trained workers from the previous Soviet Union. Spillover of tacit knowledge, the most important type of knowledge and knowledge transfer for high-tech R&D, is easily achieved in a small informal society, in which shared army high-tech training and service constitute social bases of the high-tech industry.

This workforce has been blended with the emergence of global venture capital, flexible in its places of investment. Thus, in 2000, Israel ranked second in the list of foreign companies traded in NASDAQ, and the total foreign investment of venture capital reached $3 billion, declining by about one-third in 2001, following the NASDAQ crisis and the security problems in the region (see Kipnis 2002). Domestic investment in the high-tech sector has been typified by well-developed government programs, aimed at the provision of capital and technological incubators for new ideas and firms. These programs have already been successful in retaining royalties which have been reinvested by the same governmental programs in new initiatives for the development of new ideas and technologies.

Metropolitan Tel-Aviv is the main national center of the Israeli high-tech industry, hosting almost all start-up activities. Tel-Aviv serves as the national financial center and gateway, as well as the major concentration of producer services (see Kipnis 2001). However, secondary high-tech concentrations are located in Jerusalem and Haifa, though mostly for larger R&D and production firms.

3.5 Conclusion

Information technology is critical for contemporary information production, transmission, and consumption. Its development and production is based on various types of knowledge leading to innovation. The most important source of knowledge is the personally unique accumulated knowledge, capabilities and experience of those engaged in R&D, the so-called tacit knowledge. For this type of knowledge to be productive and innovative, cross-fertilization is crucial, even beyond the boundaries of employing firms. Hence the importance of the geographical location of R&D within specific regions, which permit and foster knowledge spillover.

The geography of innovation in information technology becomes more complex, given the need for venture capital as well as additional infrastructure requirements for the industry to emerge and flourish. The increased globalization of capital flows, made possible through information technology, permit a wider global distribution of high-tech centers, though the veteran industrial cores are still dominant in the global distribution of high-tech centers. Information technology permits further the creation of global networks among these centers, something essential for centers which house plants owned by MNCs. It is, thus, for constantly innovative information technologies to permit the growing handling of information. At the same time, existing information technologies, as well as information per se, facilitate the further development of technologies. We shall return to the interaction between the geographies of information technologies and information production at the conclusion of the next chapter, devoted to the geography of information production.

The NASDAQ crisis of 2000 has brought about a slow down in venture capital investments, the fuel of the innovative industries. It has further served as a shake-out of poorly planned and funded enterprises. This was coupled with lower demand for existing products. However, this crisis has not brought about any structural change in the intensive use of information technology and information through computers at large, and through the Internet in particular. Rather, it has meant a reaction to some too-high waves of innovation cycles and profits. It may well be that this crisis may signify an integration of the new economy within the old one. The horrible terror attack on the World Trade Center in New York, one year later (September 2001), may create not just immense investments in reconstruction, but may call for new rounds of innovation in information technology, focusing on security, back-up systems, and mobile telephony. These may bring about a renewed spatial division of labor with a new round of back-office development, whether in suburbia or in other metropolitan areas altogether.

PRODUCTION

'The fact that information and content can be easily and widely distributed is often mistaken for an indication that the organization of this business is also necessarily diffused.'

(Zook 2000, p. 411).

This and the following chapters will be devoted to the geography of information as a class of communicative materials. As Abler (2000) noted regarding telecommunications, information too has to be examined separately from the perspectives of production and consumption. Whereas it might well be that the geographical *patterns* of information production and consumption may be partially similar in terms of leading cities and countries, the geographical *concepts* behind such concentrations have to be identified separately, given that the forces in operation for the production and consumption of information may differ from each other.

4.1 Information Volumes and Origins

4.1.1 How much?

A recent study performed by the School of Information Management and Systems of the University of California at Berkeley entitled *How Much Information?* attempted to estimate the volume of information production for 1999 and 2000 by type of information (Lyman and Varian 2000) (Table 4.1). Their upper estimates were based on published data, whereas their lower estimates took into account duplication and compression factors, such as hardcover and paperback books, and MP3 storage for music CDs.

Table 4.1 Worldwide production of original content, stored digitally using standard compression methods, in terabytes circa 1999

Storage medium	Type of content	Terabytes/year Upper estimate	Lower estimate	Growth rate (%)
Paper	Books	8	1	2
	Newspapers	25	2	−2
	Periodicals	12	1	2
	Office documents	195	19	2
	Subtotal:	**240**	**23**	**2**
Film	Photographs	410 000	41 000	5
	Cinema	16	16	3
	X-Rays	17 200	17 200	2
	Subtotal:	**427 216**	**58 216**	**4**
Optical	Music CDs	58	6	3
	Data CDs	3	3	2
	DVDs	22	22	100
	Subtotal:	**83**	**31**	**70**
Magnetic	Camcorder tape	300 000	300 000	5
	PC disk drives	766 000	7 660	100
	Departmental servers	460 000	161 000	100
	Enterprise servers	167 000	108 550	100
	Subtotal:	**1 693 000**	**577 210**	**55**
TOTAL:		**2 120 539**	**635 480**	**50**

Source: Lyman and Varian (2000) Table 1.

Based on their data, the total annual production of information is between 1 and 2 exabytes (one exabyte equals 1 billion gigabytes), or an average global per capita production of 250 megabytes. For comparison, 5 exabytes amount to all the words ever spoken by human beings. Magnetic storage is by far the preferred mode of information storage, comprising some 80% of information storage, which is striking when compared mainly with printed documents, comprising just 0.003% of information production! On the other hand, however, information storage, of all types, by individuals, comprises a significant portion of total information production and storage. Thus, office documents stored by individuals amount to 80% of all office documents, while photographs and X-rays are 99% of all film production (the remaining 1% is motion picture production!), and roughly 55% of all hard drives that are installed in PCs. Being most flexible and inexpensive for copying, distribution, search and mailing, Lyman and Varian (2000) note the tremendous growth of digital information storage.

Digital communication was estimated by Lyman and Varian (2000) to amount, in 2000, to some 11 379 terabytes (1 terabyte equals 1000 giga-bytes), consisting of 11 285 terabytes for 610 billion e-mail messages (a higher estimate is 1.1 trillion), 73 terabytes for Usenet, and just 21 terabytes for the 2.1 billion pages of the World Wide Web (WWW). The Web is growing by an estimated 7.3 million pages per day! The Web includes yet another 'layer', that of the *deep or invisible Web*, consisting of specialized databases and dynamic websites, not known to regular surfers. This deep Web is 400–500 times larger in volume than the *publicly indexable Web* or the *surface Web* (see also Lawrence and Giles 1999). E-mail is thus 500 times larger in volume than the surface web!

We may compare e-mail to the closest electronic communications system, namely the telephone, and the Web to the closest electronic information sources, radio and television. Lyman and Varian (2000) estimated the total volume of world annual telephone calls as 7.2 petabytes (1000 terabytes equal one petabyte), so that e-mail information volume amounts to just 0.16% of the total telephone volume! As for types of telephone traffic, data traffic was estimated to surpass phone traffic by 2002. Furthermore, the volume of fax transmission was estimated at 400 billion pages a year, totaling 6 petabytes.

The Lyman and Varian (2000) estimate for global production of radio and television programming lies between 66 and 112 petabytes, so that the web volume amounts to 0.02–0.03% of the volume of these two other electronic sources of information! Even the much larger deep web estimated at 7.5 petabytes by volume, is just 6.7–11.4% of the total volume of radio and television programs. Lyman and Varian (2000) did not provide a global estimate for surfing the Internet, other than stating that the traffic grows by 100% per year. However, they estimated that one third of U.S. telephone calls were modem calls, for both e-mail and surfing. Given the relatively low volumes of information contained in e-mail, and the relatively high volume of website loading by users through surfing, surfing seems to constitute a major 'dynamic' volume of information, notably compared to its 'static' rather small volume. Altogether, then, the Internet is weighed as a relatively small information system by volume.

4.1.2 From where?

As for major geographical foci of information production, either by country or city, data are much more scant. We shall focus later in this chapter on locational aspects of Internet information production, but it is important first to state here that the U.S. is by far the world leader in information production at large. Lyman and Varian (2000) estimated that the U.S., which is the home for some 5–6% of world population, produces some 25% of world textual information, 30% of world photographic information, 42% of world minute telephone calls, and at least 50% of information stored on magnetic media. Thus, even after three decades of evolution of the information society, the U.S. leads by far in information production of various types. Information flows in the U.S. through electronic media totaled 740 938 terabytes for 1999, divided between telephone (including the Internet) (77.7%), post (20.3%), television (1.9%), and radio (0.1%).

The share of leading global cities in information production may be demonstrated through two examples of global cities, Tokyo and New York. Lakshmanan *et al.* (2000, p. 75) used Japanese governmental data for 1987 to demonstrate the striking predominance of Tokyo in Japanese information production of all types (Table 4.2). They concluded that Tokyo alone produced some 62% of information in Japan, followed by Osaka with just 12%. On the other hand, information destined *for* Tokyo amounted to just 16% of in-flows of information in Japan, followed by Osaka with 10%. For all other regions the ratio between incoming and outgoing information is less than one. Tokyo was thus described by them as an *information gateway*. Comparing the relatively low predominance of Tokyo in telephone calls (17%) to the predominance of telephone calls in the total flow of information in the U.S. (77.7%), accentuates the special role of the telephone as the most popular and geographically dispersed medium for the transmission of both personal and business information in developed countries.

Table 4.2 Share of Japanese information production by Tokyo, 1987

Information type	Share in %
Personal	
Telephone calls	17
Computer networks	47
Facsimile transmissions	35
Postal mail	31
Mass media	
TV broadcasting	78
Newspapers by volume	40
Other publications	90

Source: Lakshmanan *et al.* (2000) p. 75.

It was estimated for New York that 75% of all telephone calls made in Manhattan were to locations within Manhattan (Moss 1986), thus attesting to the complex web of interactions within the city of New York. Generally, it was estimated that the 6% of the U.S. population residing in metropolitan New York produced, in 1994, 35% of all U.S. outgoing calls (excluding leased lines) (Graham and Marvin 1996). At the global level of electronic interactions, it was estimated for the late 1980s that New York produced more than a third of U.S. outgoing international telephone traffic, amounting to about 6–8% of the global international traffic (excluding leased lines) (O'Neill and Moss 1991, Kellerman, 1993a). Similarly, 36% of the overnight transfers to FedEx foreign centers in 1990 originated in New York (Mitchelson and Wheeler 1994). The New York pattern presents the intensive ties of the financial sector within Manhattan, on the one hand, and the strong ties of New York with foreign countries, once again probably mainly due to the financial sector.

Graham and Marvin (2001, p. 316) quote similar trends in other global centers as well. In France, some three-quarters of advanced data traffic are generated in the Paris region, and in London, terminate some 55% of all U.K. terminating international private telecommunications circuits.

The data for the U.S., Tokyo, New York, London, and Paris suggest that despite the potential dispersal power of electronic information, both its production and consumption are unevenly produced, not just along the *digital divide* between developed and developing countries (see Chapter 7), but also among countries in the developed world, as well as among cities, even in the leading country, the U.S. In the following sections (as well as in Chapter 8 for consumption) we shall attempt to highlight the reasons for this concentration.

4.2   The Internet: Evolution and Structure

4.2.1   Essence

One way of looking at the Internet is through the eyes of computer networks: 'The Internet consists of a global network of computers that are linked together by 'wires' – telecommunications technologies (cables of copper, coaxial, glass, as well as radio and microwaves)' (Dodge and Kitchin 2001, p. 2). Another way of looking at the Internet is from the perspective of information technology, namely that it consists of computers interconnected by the protocol TCP/IP (Transmission Control Protocol/Internet Protocol)

(Dodge and Kitchin 2001, p. 2). Yet another way of viewing the Internet and the similar intra-organizational Intranet systems is through the form of information exchanges, namely that they constitute networks which permit open and free information exchanges (see Nolan 2000).

For our purposes, the Internet, and specifically the Web, constitute *the most comprehensive information system*, when compared to any other information systems producing or transferring much larger volumes of information. Furthermore, at the same time, the Internet also constitutes a communications system. The Web contains all the information forms developed so far, whether it be text, data, graphics, voice, or motion pictures. Unlike the telephone it contains structured information, and unlike almost all radio and television programs it is an on-demand system. Furthermore, the Internet is an interactive information system, thus permitting the performance of a variety of economic activities, such as shopping, ordering, transfers, etc. In fact, it permits through the Web, as well as through e-mailing, an incorporation within it of all other electronic and printed forms of information, e.g. fax, voice telephony, radio, television, newspapers, and books. Through new generations of mobile telephony technology the Internet is gradually becoming a completely wireless information system. As such, the Internet may be considered a *general purpose technology* (GPT), in a similar way to electricity, railroads, and automobiles (Harris 1998; see also Malecki and Gorman 2001).

Another way of looking at the Web is as a globally accessible library containing information on any field of human interest. From a geographical perspective, traditional libraries are typified by their concentrated location, in that a library building is supposed to store as many books and journals as possible, in order for it to permit an ultimate ease of access by the readers and/or librarians, who still have to move from one location to another within a library complex. The Web, on the other hand, is extremely distributed, since websites may be hosted on server farms and hosting computers in various parts of the world, but still permitting instant electronic access by readers and information specialists, stationed each at just one computer for any required information item.

Furthermore, this electronic library is typified by an almost absolute fluidity and flexibility of its contents, given the ease of adding information materials to the system by numerous and geographically distributed website patrons. This fluidity is, though, different from traditional libraries, in a negative aspect, namely the frequent deletion of information and of whole websites from the system, so that information storage on the Web is not stable, as far as previously existing materials is concerned. This stems from the different system of information ownership between these two forms of libraries. A traditional library is an organized and professional institution priding itself on the continuous richness of its collection, so that the removal of an item from the collection is a librarian's professional decision, and the physical packages of information items, whether they be books or journals, are owned by the library. Web information is owned by a huge number of website/domain owners, who may remove materials or delete whole websites for a variety of reasons. On the other hand, the Web constitutes many times the most updated library, given the ease of adding materials to it. From a legal perspective, the Internet at large, and the Web in particular, are viewed much more widely. In several legal proceedings in U.S. courts the Internet has been variously described or defined as a library, a telephone, a street corner or a park, broadcast media, and a shopping mall (Biegel 2001, p. 27).

4.2.2 Evolution

The Internet is currently an information and communications system available to the general public and geared to the needs of the widest possible customer sectors. Originally, however, it was an outgrowth of the communications and information needs of two sectors, namely the U.S. military and academic communities. The U.S. army was interested back in the 1960s in a back-up computer-based communications system, in case the telephone system were to be severely damaged by a nuclear attack. The early connections of the military system have been made to university research centers, so that the new communications system, later known as e-mail, has brought about the development of academic networks, connecting universities and research centers. Side by side with the emergence of an electronic inter-computer communications system, were developed information technologies which were needed for the establishment of an *information system*, and which later permitted the formation, storage, transmission and retrieval of data, information, and knowledge bases, first known as Gopher and FTP (File Transfer Protocol) later transformed into the WWW (World Wide Web). This system first served the specific needs of universities and research centers.

Early experiments with an inter-computer communications system were performed by the U.S. army in the early 1960s. These experiments became feasible with the manufacturing of the first all-electronic telephone switches in 1962, and the invention of packet switching, permitting the switching of data traffic, in 1964. These developments followed the rapid developments in computers and telecommunications in the 1950s (see Kellerman 1993a).

PRODUCTION

The experimental ARPANET (Advanced Research Project Network) system was launched by the U.S. army in 1969 connecting to a site at the University of California, Los Angeles (see e.g. Townsend, 2001a, Nolan 2000, Warf 2001, and Dodge and Kitchin 2001, for reviews). The geographical diffusion and expansion of the system was aided by the development of the TCP/IP communications protocol in 1973, permitting the transfer of electronic information between computers with different hardware, operating systems, and applications software.

The 1980s witnessed three major developments spanning over the U.S. and Europe: the maturing and diversification of computer communications networks; further developments of communications technologies; and the development of technologies for the establishment of the Web. Thus, in 1987 the original military ARPANET system was replaced by the civilian and educational NSFNET (National Science Foundation Network) system, based on a U.S. transcontinental backbone of communications lines and hubs. In 1981 Yale University and CUNY (City University of New York) were connected into what became another inter-university system called BITNET (Because It's Time Network) in North America. As of 1984 it was coupled with IBM-sponsored EARN (European Academic and Research Network) (Kellerman 1986). These systems were based on the chain principle, so that each new linking university was connected to an already connected one, while permitting other joining institutions to connect through its computers. In 1995 the Clinton–Gore administration decommissioned and privatized NSFNET into what became the Internet, a unified non-governmental, general-purpose, global computer communications and information system, a *network of networks*.

The major European research center CERN, based in Switzerland, developed in the 1980s through the early 1990s, some major information identification and switching tools for the emergence of the Web: HTML; URL; and HTTP. At the same time, in 1986, the American Cisco company introduced the router, permitting route selection for the transmission of information between computers. Meanwhile, at the American university R&D frontier, UNIX was developed at the University of California, Berkeley, and, in 1993, the first browser Mosaic, was developed and introduced at the University of Illinois, Champaign-Urbana. This latter tool marked the beginnings of the Web as a collection of sites for user downloading. The Mosaic browser served as the basis for the introduction of *Netscape* (1994) and *Explorer* (1995). Thus, in 1994–1995, some 26 years after it was experimentally launched, the Internet was established as an open, commercial communications system, as well as rapidly becoming a comprehensive information system.

From a geographical perspective, it is interesting to compare the diffusion and expansion of the original military ARPANET/NSFNET system with the academic BITNET system across the U.S. (see Kellerman 1986, Townsend 2001a). Being oriented to military centers and related research centers, ARPANET did not diffuse necessarily through the major population centers. Thus, in 1971, Los Angeles, San Francisco and Boston were connected through several sites in each city, but New York had only one node in New York University. This trend continued also when the backbone was established by NSFNET in the late 1980s, resulting in highly networked urban regions in Seattle, Austin, and Boston. On the contrary, BITNET, as an academic network originating in New York, had the largest number of universities connected in New York, followed by California (Los Angeles, San Francisco), and later on by Massachusetts (Boston), Illinois (Chicago), and Texas (Austin). Given the chain structure of the network, the diffusion of BITNET was both contagious and hierarchical. Interestingly, the major nodes on both NSFNET and BITNET emerged in the 1980s as the major centers of information technology R&D and production. As we shall see in this and in later chapters, the early lead in information technology R&D has gone hand in hand with evolving patterns of information production and consumption.

4.2.3 Structure

The structure of the Internet may be highlighted from a geographical perspective through three aspects. First, the rather fixed transmission system, consisting of nodes, hubs, and lines (see Chapter 6); second, the dynamic flow patterns of information within this system (see Chapter 8); and, third, the addressing or identification system of information producers and consumers, which will be briefly discussed below.

The *domain name system* (DNS) is the system which identifies any hosting server by a numerical address, the *IP address* and its equivalent alphanumerical address (Wilson 2001a). Responsibility for allocation and registration of IP addresses and domain names was, first, exclusively located in the U.S., moving from governmental agencies through the NSF to a non-profit organization, *Network Solutions*. As of the late 1990s these activities have also been maintained by a European company, *NetNames*. In January 2001 the share of *Network Solutions* in world domain registrations was 52%, down from 92% in just one year. In January 2001 the world total for domain name count was 33 million (Zook 2001a).

Top level domains (TLDs) are the suffix of any e-mail and Web address. In the U.S. this suffix constitutes a three or four letter code of one of ten categories, the generic TLDs (gTLDs), the three latter of which were added in 2001–2002:

com–commercial company
org–non-profit organization
net–network
edu–academic institute
int–international organization
gov–governmental department/agency
mil–U.S. military
name–personal (not yet operational in early 2002)
biz–business
info–product/organization information.

Some less frequently used gTLDs include aero, museum, and pro. For June 2001, Zook (2001a) reported that some 76% of the top four TLDs (CONE: com, org, net, edu) registered domain names worldwide were .com, some additional 9.4% were .org, close to 15% were .net, and some 22% were .edu. The total share of the U.S. in world domain names has been declining in recent years, reaching 42% in July 2000, remaining such through January 2001 (down from 50% a year earlier) (Zook 2001a).

Outside the U.S., the TLDs consist of a two-letter code representing one of about 250 countries, the ccTLD, normally the common international country code, such as uk for the United Kingdom. This code is preceded by a two or three letter code representing gTLDs, such as co for commercial companies, ac for academic institutions, and org for organizations. The United Kingdom is second to the U.S. in the number of domain names. The ccTLD uk for the United Kingdom and gTLDs registered in the U.K. accounted for 12.4% of world domain names in January 2001 (13.1% in July 2000, up from 8.5% in January 2000). Germany was third with the country code de and German gTLDs took a share of 10.3% of world domain names in January 2001 (9.3% in July 2000, and 8.6% in January 2000) (Zook 2001a). All other countries had percentages of lower than five, so that some two-thirds of world domain names are registered in the three leading countries. The decline in U.S. hegemony, which leveled off in the second half of 2000, was, thus, being coupled with a rising centrality of two European nations, only one of them English speaking.

The TLD is preceded in a domain name by the company/organization full name or its abbreviation, or any other name that the registering company/organization would like to carry for its identification on the Internet, if still available. This domain name may be preceded by a local host or server name or any sub-domain. For example, the URL 'exchange.haifa.ac.il' refers to the server Exchange at the University of Haifa in Israel.

Domain names are traded, and registering entities may sometimes choose registration countries for various convenience factors (Wilson 2001a). Thus, there does not exist a full symmetry between a company/organization country of operation and country of Internet/host registration (see Zook 2001a). By the same token, within a given country, the Internet registration address of an entity is not necessarily its postal address or major address of operation.

4.3 A Conceptual Framework for Information Production

4.3.1 Process

The very concepts of *information production* or *content business*, vis a vis the Internet industry, seem somehow confusing if they refer to the process of creation of information and not to *informational products* (see Zook 2000). Information per se may obviously be produced independently of the Internet by individuals, organizations and businesses, and this information may be processed and distributed through many traditional or electronic media, other than the Internet, such as books, magazines, or television. Thus, the so-called Internet industry is in charge of packaging and processing information for the Web, followed by its distribution through computer networks (Figure 4.1). The industry adds high-tech *expertise* or *knowledge* to *information* produced and provided by individuals, organizations, and businesses, in order for it to be processed and packaged for Internet transmission.

The relationship between information producers and Internet packaging and processing knowledge-expertise is similar to the one between authors and the traditional, and until recently, low-tech, book publishing industry. We may thus follow the model for this latter industry, proposed by Locksley (1992), who differentiated among four *quadrants*: *producer* (author); *packager* (publisher); *distributor* (wholesale/retail); and *user* (reader) (see also Goodchild 2001, p. 73). For the Internet industry all four quadrants apply as well, plus an additional one, namely domain registration, which is to some degree similar to the formulation of a specific ISBN (International Standard Book Number) for each new book. ISBNs, like domain names, contain the publishing company code, and thus its address. On the other hand, domain registration implies the establishment of real and virtual addresses for a

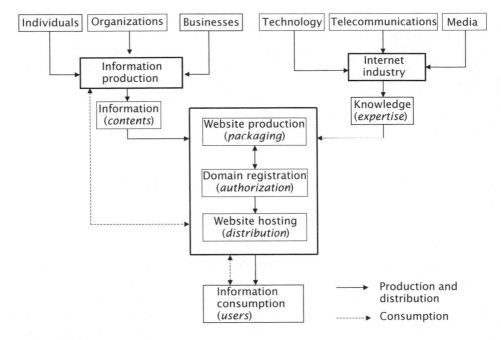

Figure 4.1 Phases in information processing

new website, relating, however, to the information producer rather than the packager. We may therefore identify, the following phases for information production and processing for and through the Internet: *producer* (individual, organization, business); *packager* (website producer); *authorizer* (domain registration agency/company); *distributor* (website hosting computer); *user* (Internet customer).

4.3.2 Location

From a locational perspective, information production has been viewed as deeply embedded in urbanism, and even more so as it is strongly related to growing globalization (see e.g. Sassen 1997). Information producers may thus be located almost anywhere. One may assume, though, that the volume of information production by places is a function of population size, as well as of economic specialization, with the finance industry and related insurance and real estate sectors (FIRE) being heavier information producers than industrial plants. Thus, Zook (2000) found an *r-square* ratio of

0.65 in a regression analysis of U.S. SMAs, in which the number of .com domains served as the dependent variable, and population as the independent one. If the production of information had been just a function of population size, then the regression coefficient should have been much higher.

The location of the Internet industry, notably the packaging (website production) stage, is more concentrated. 'Cities are driving, rather than simply participating in, information-based phenomena such as the expansion of the Internet' (Moss and Townsend 1997, p. 47). This driving is unequally distributed among and within cities. The industry tends to be located in major cities, and in central areas within cities where relevant workers are concentrated, and where daily contacts may be maintained (see Braczyk *et al.* 1999). Certain economic specializations of specific cities may bring about leadership in Internet production in that city. Such leadership in the production of Internet sites may rise out of a city's previous specialization in finances, high-tech, and media (Kellerman 2000a).

The interrelationship between finances and information in general, and between global finances and telecommunications in particular, has long been very intense. Historically, 'the ability to collect, exchange, rearrange and interpret *information* is the most persistent characteristic of an international finance center' (Laulajainen 1998, p. 257). As for the adoption of telecommunications by the finances industry: 'there is little doubt that the electronic integration of (these) financial and capital markets represents the single most important application of telematics within the global economy' (Graham and Marvin 1996, p. 144).

The extensive reliance of contemporary banking on telecommunications could have brought about a globally very dispersed financial industry. However, global financial markets in the era of telecommunications have rather become more concentrated, under the leadership of New York, London, and Tokyo, simultaneously combined with a globalization of customers. This concentration has been attributed to external economies of scale, demand for skilled manpower, need for face-to-face communications, and the ability to liquidate instantly (Brealey and Ireland, 1993). A global financial center, equipped with a dense and sophisticated telecommunications infrastructure, as well as the local existence of financial tradition and manpower experienced in information handling, may yield a strong specialty in information production and packaging as independent industries.

The second economic specialization which may lead to specialization in Internet production is the high-tech industry. This industry too is a heavy user of telecommunications, given the very nature of its activity and the geographical spread of R&D, and particularly of customers. However, the requirements of high-tech industries for a sophisticated telecommunications infrastructure are lower than those of global financial centers. The major contribution of a locally concentrated high-tech industry to a specialization in Internet production lies in its manpower, as well as in the immediate local availability of Internet-related technological innovations, thus providing for a local leading edge in a constantly dynamic industry.

The third local economic specialization which may lead to specialization in Internet production is the media, notably visual ones, such as television, motion pictures, arts, etc., as well as the areas of advertising and multimedia. These economic activities may provide the expertise required for the design and structure of Internet websites. The emergence of the Internet industry has revived demand for writers and artists who prefer living in older neighborhoods of large cities (Kotkin 2000, pp. 20, 120, Scott 2001).

Zook (2000) argues for a possible strong geographical fit between the location of content production and the address provided in the domain registration procedure:

1. *Registration of a .com domain name indicates a higher degree of purposeful and commercial information distribution than just surfing or e-mailing friends and family.*
2. *There is no inherent geographic bias in the means of registering a .com domain name.*
3. *The registration address, particularly for newly-conceived Internet ventures, corresponds to the location of the site where content location is taking place.*
4. *It is a marketing and brandname necessity for Internet contents firms to have their own .com domain name. (p. 413).*

At the empirical level, Zook (2001a) reported an analysis of *CorpTech* data consisting of the operational real addresses of 50 000 high-tech companies showing a high fit of 73–84%, depending on the level of geographical fineness of analysis, between these addresses and the ones reported as the domain name addresses. It is, though, questionable, whether addresses of commercial companies in areas other than computers would show the same levels of fit. Furthermore, it is not known from the address of information production (company operational real address) and the one registered for the domain, where the information was packaged (where the website was actually produced).

The last phase of Internet information production and distribution, before reaching customers, is the hosting computer, the machine in charge of distribution of the information to users. We shall discuss the geography of this phase in Chapter 6, but it would suffice to note here that hosting computers do not have to be located either in the place of contents production, or in the places of website production or domain registration, as the efficiency of transmission may be the major selection criterion for hosting computers (Zook 2000).

The convergence, in certain cities, of all the components of the Internet industry would turn these cities into leaders of the information economy. These components include: the production of information (content) through concentrations of specific economic activities; the availability of the knowledge required for the Internet industry (high-tech and media), and a local existence of a strong telecommunications infrastructure for both the facilitation of Internet expertise, as well as for transmission capacity. In the next sections we will attempt to identify these cities.

4.4 Ranking Urban Centers of Information Production

By their very nature cities serve as centers for information production, in their capacity as coordination, control and command centers for various economic activities, at varying geographical scales of influence. Cities further serve as networks for people and organizations, as well as foci for cultural activities. In addition, cities produce and transmit information through mass media at the local, regional, national and international levels (see e.g. Castells 1989, Short and Kim 1999, Graham and Marvin 2001). Hall (1997) noted that ranking cities by their production of more sophisticated information, has to be 'in relation to the informational economy. Their position depends on their capacity to generate, process and exchange information, especially higher-order information of a somewhat specialized and privileged kind' (p. 318).

Measuring the production of Internet information is not possible by direct measurement, rather the number of domains per city has been used, and this constitutes a crude measure. One may assume that any domain represents at least one Internet website, if not more. It is, though, difficult to determine the number of sites per domain, as well as the size of the sites by page numbers, which may differ widely among domains. Also, one organization may register several domains. Furthermore, it is assumed that the billing

95

address of domains represents their operational location, and that the place of information transmission is also the place of information or website production. All these assumptions and reservations may apply to many or just a few domains, but still the identification of leading urban centers by the number of domain names may attest to a concentration of Web information production.

4.4.1 A global view

From a global perspective, the urban concentration of domain name leadership is even more striking than the national one, with about 25% of world domains located in just 10 cities, containing 1.5% of the world population in January 2001 (Zook 2001d) (Table 4.3). Just four cities, New York, Los Angeles, London, and San Francisco contained about 17% of world domains, with a sharp decline between the fourth city (San Francisco), and the fifth one (Washington, D.C.). New York and Los Angeles each contained close to 5% of world domains. London is high on the list because of the concentration of British domain registrations in this city, as we shall see below. For this same reason, no German city is on the list, despite Germany's high ranking in domain name registration. As we shall see, domain registration in Germany is rather dispersed among several leading cities. The only additional non-American city among the top ten was Seoul. We shall see in Chapter 9 the leadership of Korea in advanced broadband Internet consumption, which may have reflected on the supply side of the domestic Internet economy,

Table 4.3 World leading cities in the number of domains, January 2001

City	Domains (gTLD & ccTLD)	% of world total
New York	1 575 500	4.9
Los Angeles	1 463 900	4.6
London, U.K.	1 182 928	3.7
San Francisco	1 010 550	3.1
Washington, D.C.	642 250	2.0
Seoul, Korea	560 796	1.7
Chicago	475 800	1.5
Boston	457 600	1.4
Miami	340 500	1.1
Dallas	318 750	1.0

Source: Zook (2001d).

through the large number of domains registered in Seoul. As we shall see below, in Korea the concentration of domain registration in Seoul is very high, so that the advanced Korean Internet system and the urban geography of the Internet production industry have brought about the prominence of Seoul at the global level. Seoul's prominence is impressive also in light of the leadership of London, the domains of which are in English, the most popular language on the Internet.

From a wider urban perspective the distribution of domain registrations is more deconcentrated. As Zook (2001a) reported for January 2001, the 100 leading cities worldwide in domain registration accounted for more than half of domain name registrations, Fifty-seven of these were located outside the U.S. compared with 45 in 1998. Back in January 1999, of the twenty world leading cities, 17 were in the U.S. (London ranked 4th; Toronto 17th; Tokyo 20th) (Zook 2001b).

4.4.2 The U.S.

The data in Table 4.4 for 2000 show a concentration of 42.8% of U.S. domains in just nine metropolitan areas. Moreover, the first three cities, New York, Los Angeles, and San Francisco serve as the location for over a quarter of U.S. domain names (27.4%). This concentration is not a new phenomenon, but a trend that has prevailed since the emergence of the Web in the mid-1990s, with Los Angeles joining as a leader in 1996 and replacing San Francisco as second in 1999 (see Moss and Townsend 1997, 1998, Zook 1998). Thus, despite the global decentralization of domain names, from the U.S. outwards, the urban concentration within the U.S. remains rather strong, and as Graham and Marvin (2001, p. 329) suggested, it is even growing. However, the diffusion of Internet domains to other countries has reduced the world share of New York as the global leader of world domains, from 5.9% in January 2000 to just 4.9% a year later (Zook 2001a).

As for the population rankings of the leading U.S. cities in Internet information production, with the exception of the two globally and nationally leading cities, New York and Los Angeles, the larger population of a city does not necessarily guarantee a similarly sized information industry. There are still rapid changes in the rankings of cities below the six leading ones, as is demonstrated by the July 2000 data (Table 4.4) compared with those of January 2001 (Table 4.3). For most cities in Table 4.4, specialization in high-tech/information technology, or knowledge industries, may provide a clue for their specialization in information production or packaging. San

Table 4.4 Domains in U.S. metropolitan areas by % of national total, July 2000

MSA	% domains	Population rank
New York (NY; CT; NJ)	12.4	1
Los Angeles-Long Beach-Orange County	9.7	2
San Francisco-San Jose	5.3	5
Washington (DC; MD; VA; WV)	3.8	4
Chicago	3.3	3
Boston (MA; NH)	2.5	7
Atlanta	2.0	10
Philadelphia (PA; NJ)	1.9	6
Seattle-Bellevue-Everett	1.9	11
Total	**42.8**	

Sources: Domains (Zook 2001a); population (Townsend 2001a).

Francisco/San Jose is a leading example in this regard, enjoying also a well-developed media industry, concentrated mainly in the *Multimedia Gulch* (see Table 3.5). A match between the addresses of registered domains and the locations of business operations of the website-owning firms, may indicate either a large number of websites dealing with high-tech, or it may present the local existence of Internet expertise, which may lead to the development of an information industry dealing with the production of websites in areas other than high-tech. We shall return to this question in the next chapter, which deals with contents.

Washington is one of the leading cities in information production. It is the home not only of security-related R&D, but also of extensive Federal government Internet services, as well as the home of AOL *(America On Line)* and *Network Solutions* headquarters (Moss and Townsend 2000). Furthermore, the Washington metropolitan area enjoys the highest per capita income, as well as the highest proportion of college graduates in the U.S. (Townsend 2001a), which may reflect on both the production and consumption of information.

4.4.3 Europe

As mentioned earlier, the U.K. and Germany follow the U.S. in the percentage of world domain names. However, the urban distribution of domain name addresses is rather different from the American pattern, at least as far as the leading urban centers are concerned.

London is a non-U.S. city leading in information production. Like New York, it is a global financial center, but unlike Tokyo, the third pole in the top of the global financial system, it is an English speaking city, with English still being the predominant language on the Internet. London is blessed with the three major factors mentioned earlier for the emergence of a leading Internet industry: its financial sector is tied into an excellent telecommunications system, and the city is also a leading high-tech center (see Table 3.3). Furthermore, London has a highly developed media industry, well connected with Hollywood, via the *Sohonet* (Graham and Marvin 2001, pp. 332–333).

In July 2000 the share of London in British domain registration was 22.1%, compared to 12.4%, the share of New York in American domain registrations at the same time. This high share of London in British domain registrations reflects the global leadership of London on the one hand, and the much smaller size of the U.K. compared to the U.S., where even global leadership is divided between two cities, New York and Los Angeles, plus additional leading cities, notably San Francisco. Similar patterns of primate city concentration were shown for other European cities and countries as well. Thus, Paris accounted for 26.5% of French domains, and in Spain, Madrid and Barcelona together represented over 50% of the Spanish domain name registrations (Castells 2001, p. 220).

Table 4.5 presents the distribution of domain registration among major German cities, similar to Table 4.4 for U.S. cities. As we have mentioned already, Germany has become the third leading country in domain registration, and first among the non-English speaking countries, possibly reflecting its economic might. Since no German city has a global financial status such as those of London or New York, and since websites in German serve mostly Germans and German speaking nations, the distribution of domain registrations among cities in Germany is much less concentrated than in the U.S. and the U.K. In fact, the percentage of domain names registered in the leading city in Germany, Berlin, is similar to that of San Francisco, ranked third in the U.S., given that New York and Los Angeles are rather global centers for information production. Also, the total share of the nine leading German centers is about half that of the equivalent ones in the U.S. (23.2% and 42.8% respectively). In Germany, as in the U.S., there is no match between population ranking and the share of domain names, and again this is notable especially at the ranks following the leading centers. Thus, there are cities

Table 4.5 Domains in German metropolitan areas by % of national total, 2000

Urban area	% domains	Population rank
Berlin	5.8	1
Munich	4.3	3
Hamburg	4.0	2
Cologne	2.5	4
Düsseldorf	1.7	9
Frankfurt	1.7	5
Stuttgart	1.2	8
Bonn	1.0	20
Hannover	1.0	12
Total	**23.2**	

Source: computed from DENIC eG (2001).

which specialize in information production. An interesting example in this regard is Bonn, which is ranked 20th in population size, and 8th in the number of domain names, reflecting its status as previous capital city of West Germany. However, whereas the leading centers in the U.S. are located on the two coasts as well as in the inner part of the country, in Germany all the leading centers are in the former West Germany, attesting to the economic gap between the two parts of the country (see also map in DENIC eG, 2001).

4.4.4 Asia

The Korean distribution of domain name registration among cities looks very different from both the American and the European ones. In 2001 the share of Seoul was 59.8%, down from 61.8% a year earlier. The share of the Capital Region in 2001 was 76.1% (down from 77.4% in 2000). This extreme concentration was explained by the status of Korea as a developing country dominated historically by a primate city (Huh and Kim 2001). However, with regard to Internet penetration and its advanced use, as well as the number of domain registrations, Korea is more of a leading country (see Chapter 9). Thus, the Korean pattern might present either an example of an Asian pattern, or a pattern reflecting fast industrializing countries (*tigers*).

The Chinese pattern is similar to the Korean one (Table 4.6). The capital Beijing led in Internet production in 2001, though with a lower percentage compared to Seoul (35.3). The leadership of the capital city, Beijing, is striking despite its being second to Shanghai in terms of population size, and

Table 4.6 Domains in Chinese metropolitan areas by % of national total, 2001

Urban area/Province	% domains
Beijing (1)	35.3
Guangdong (2)	14.4
Shanghai (1)	9.5
Jiangsu (2)	5.2
Zhejiang (2)	4.0
Total	**68.4**

(1) Urban area.
(2) Province.
Source: Lou (2001).

this may reflect the role of government, which has recently rather declined in terms of domain names. The concentration of Internet production in China in just a handful of cities has a wider dimension, since the leading cities and provinces are in the coastal and more developed areas, similar to the German pattern (Lou 2001).

4.5 Global Centers: New York and Los Angeles

4.5.1 Global centrality

Global centrality of cities may emerge in several areas (Kellerman 2002b). In the well-known arena of the globalization of the capital industry, the most important global capital centers have been mentioned already as the triad New York-London-Tokyo, representing respectively three global cores (North America, Europe and the Pacific Rim, respectively). These three global cities present a dense network of interconnections, side by side with local specializations in the global capital industry (see e.g. Sassen 1991).

Within the U.S., New York, Los Angeles, and Chicago have been recently portrayed as America's global cities, but without attention to the role of New York and Los Angeles as global leaders in information production (Abu-Lughod 1999). Contrary to the rather global dispersion of centers of global capital flows in all three economic cores, the two global leaders in information production: New York and Los Angeles, are both major American cities, one of which, New York, constitutes simultaneously a global leader for both finances and information, whereas the second one, Los Angeles, serves as a global center for the media industry.

Information also plays a dominant role in the recently flourishing *cultural products industries* (Scott 2001), or the *cultural industrial complex* (Kotkin 2000, p. 130), consisting of the media, fashion, cultural services, and the creative professions (such as architecture) (Scott 2001). By their very nature, all these activities include a strong or dominating information component. New York and Los Angeles lead in these constantly refreshing industries, attracting highly-creative professionals. Some 14% of American artists live in these two cities which constitute home for just 3% of the total American population (Kotkin 2000, p. 131). These industries tend to concentrate in specific areas within cities, notably in or close to CBDs (central business districts). They further bring about the development of global cultural symbols associated with the hosting urban entities, a process which is striking, particularly with regard to Hollywood. The development and transmission of such images are connected with a city's broadcasting and communications systems, and a city is also a command and control center for major MNCs, which can handle the promotions and transmissions of such images (Scott 2001).

Side by side with the fusion of information forms, capital was turned into information through the emergence of electronic transmission of funds. This change in global financial business is of particular significance in the development of information space, since it gives an advantage in the production of information to financial centers which existed already in the pre-electronic era. Besides the processing of capital per se, major financial centers interpret and analyze market trends, something which Sassen (1999) termed *financial engineering*. Financial centers do not only produce, transmit and receive huge amounts of information in the form of capital and interpretations of capital, but they develop a large FIRE (finance, insurance, and real estate) sector, which implies the development of additional information business in areas such as advertising and marketing. FIRE industries and information technology have mutually reinforced each other in cities which have developed into major trade centers, notably in their core areas. Los Angeles serves as a notable exception in its metropolitan decentralization of information industries (Castells 1989).

Global leadership in capital formation, processing and exchange, does not automatically imply leadership in the global information space of production. On the one hand, there might exist barriers, such as language, which may deter a financial center from becoming a global information center (e.g. Tokyo). On the other hand, there are information forms which may be produced independently of financial centrality, such as television, motion pictures, and the Internet. However, a coinciding concentration of major information production in all of these areas in one city may bring it to global information leadership. By the same token, and as will be shown later, places of peaking production of information do not necessarily coincide with places of peaking information consumption, or peaking information hardware or software production. Thus, global information spaces are complex as far as local peaks are concerned. However, the growing importance of information per se, brings about a special significance to places which lead in information production.

4.5.2 New York

New York was termed 'the information capital of the world economy' (Mitchelson and Wheeler 1994, p. 97). Note the suggested information supremacy of New York in the world *economy*, rather than in the world *culture*, a status which New York shared with other cities, notably Los Angeles. New York leads in the production of business-related information, in the widest sense of the term *business*, whether it be financial information, radio and television news, books and magazines, or Internet websites. New York has been able to keep its leadership through continuously changing, updating, and trend-setting the information types being fostered within it (Moss 2000). This amounts to a continued process of regional learning, in both the knowledge and expertise required for the facilitation and packaging of information production, as well as in information production per se.

New York gained an initial advantage as an international gateway to North America, by serving as a main entrance for European settlers and merchants, making use of boats as the major vehicle for transporting people and commodities (see Nijman 2000). This continued leadership has paved the road for leadership in finances, command and control activities, and later on in business information, using computers and telecommunications as vehicles and channels, mainly for exporting rather than importing information. New York's leadership in American FIRE business was already reached in the early 19th century. Thus, the *New York Stock Exchange* overtook Philadelphia's in 1816, followed by national leadership in banking capital and insurance in 1824 (Abu-Lughod 1999). New York enjoyed the advantage of being the business capital of a big country, compared to other world cities which were backed by smaller national territories and populations (Beaverstock *et al.* 2000).

New York, with its two leading stock markets, is the largest equity trading market place in the world, and equity trading is New York's specialization within the global triad of leading capital markets. New York was further a close second to London in 1998 in institutional equity holdings, reaching over $2 billion (*Thomson Financial* 1999), but its leadership in the global economy has been in the innovation of financial products (Sassen 1999). The New York area economy in 1990 was just 14% larger than that of the Los Angeles one, but the New York FIRE sector produced almost three times as much income as that of Los Angeles (Finney 1998). This concentration of finances and financial services is the major contributor to the world's heaviest telephone traffic in New York, mentioned at the beginning of this chapter.

As a leading command and control center, New York housed, in the early 1990s, 40 of the 100 top American multinationals, which generated 55% of the total foreign revenues created by these 100 companies (Mitchelson and Wheeler 1994). New York served also as the headquarters location of 12 of the world's leading corporations in 1997 (second to Tokyo) (Short and Kim 1999, based on the 1998 *Fortune 100* list), much more than the following American city, Detroit, with just four corporations. However, as Sassen (1991) claimed, this ranking has lost some of its importance in favor of the concentration of business and producer services, in which New York is leading as well (Short and Kim 1999).

The city has furthermore the largest concentration of book and magazine publishing, as well as being the 'global capital of the multimedia, or new media, industry' (Pavlik 1999, p. 82). However, the exact definition of multimedia is questionable, since we shall note later on the supremacy of Los Angeles and California at large, in the multimedia industries. New York is the U.S. national center for radio, TV and CATV broadcasting, though not of film production, other than news (Moss 1996). Some 30% of all Americans employed in the book and magazine publishing industry in the late 1990s worked in the New York area (Kotkin 2000, p. 132). As was found in a study of the British magazine publishing industry, editorial offices need to be located in centers of financial, business, as well as cultural activities, such as London or New York, given their need for information networks consisting of personal contacts and meetings, whereas printing, or the manufacturing of magazines, may be performed elsewhere, using telecommunications for the transmission of contents (Driver and Gillespie 1993).

Several initial cumulative advantages have facilitated the development of the largest Internet industry in the city. The supremacy of the city in finances as an information producing sector, and the heavy telecommunications infrastructure and computing experts serving this sector have been two important factors. Also, the leadership of the city's media, art, and publishing industry facilitated the global leadership of New York, as well as that of Los Angeles, in Web information production.

New York's information activities are most heavily concentrated in Manhattan, notably in its downtown and midtown sections (generally along 5th Ave. up to 45th St., the area which has been called *Silicon Alley* since the emergence of the Internet industry (see Zook, 2000, Heydebrand 1999)). This is probably the most intensive information production area worldwide. The area has enjoyed cumulative advantages since the early 19th century, with historically changing technologies for information production, processing and transfer (Moss 2000). In 1997, 4.2% of U.S. commercial domains were located in Manhattan alone, twice as many as in the second leading center, San Francisco (Moss and Townsend 1997). The dramatic growth in the so-called *.com economy* in the late 1990s, brought about rising rents for both business and residence in *Silicon Alley*, driving out less affluent occupants. Similar trends developed in San Francisco's *Multimedia Gulch*. This process has changed and was even reversed with the *NASDAQ* crisis of April 2000.

New York's leadership in the production of information has not been accompanied by supremacy in information consumption, at least not in the information types in the production of which the city has been leading (see also Chapter 8). Furthermore, New York does not lead in the production of the hardware and software required for the production of the information types led by the city. It should be noted, however, that supremacy in information production may imply political, social and economic power and importance, much more than supremacies in information consumption and information hardware and software production. The nature of the global economy may call for a more dispersed or different geographical pattern of both the consumption of information and the industrial production of information hardware and software. Mitchelson and Wheeler (1994) in their study of Federal Express data noted that New York 'talked' more than it 'listened', and 'listening' or consumption of information, notably an imported one, is determined, among other things, by socioeconomic levels of city residents. Thus, New York ranked second to Los Angeles in book sales in 1992, and third to Chicago in the number of bookstores (Table 4.7), and it ranked only eighth in Internet penetration rates in 1999 (see Kellerman 2000a and Table 8.1).

Furthermore, New York does not present any leadership in the manufacturing of the two major devices needed for the contemporary production of

Table 4.7 Top U.S. book markets, 1992

Urban area	No. of establishments	Sales ($1000)
Los Angeles-Long Beach	454	383 902
New York	363	359 716
Chicago	366	271 867
Boston	249	238 116
Washington	296	234 589
Philadelphia	268	161 736
San Francisco	167	138 875
Seattle	176	131 679
San Jose	99	125 015
San Diego	175	115 419
Detroit	171	115 319
Atlanta	171	112 850
Oakland	137	108 453
Houston	143	108 449
Dallas	146	106 771

Source: Top U.S. Book Markets (2000).

information, namely software and hardware. The software industry is mainly an American one, but it was shown that for 1986, New York State ranked low as far as the relative concentration of the software industry was concerned. This was true also for California, which led, though, in the absolute number of workers in this industry. Leading states were Virginia, Maryland, Colorado, Massachusetts and Michigan, some of which lead in high-tech industries which specialize in R&D, as well as in the production of hardware (Haug 1991). The production of PCs (personal computers) is also mostly an American or an American-owned industry, though various components are produced elsewhere (Angel and Engstrom 1995). U.S. production has been led by major high-tech concentrations, notably Silicon Valley, Southern California, Research Triangle Park, and Austin, without any concentration in New York.

4.5.3 Los Angeles

Los Angeles has been termed 'the world's leading manufactuary of global culture and popular entertainment' (Soja 2000, p. 226). More particularly, Hollywood was called 'the world capital of filmed entertainment' (Storper 1997, p. 83). Thus, 55.6% of U.S. employment in motion picture production and services in 1998 was concentrated in the Los Angeles-Long Beach area (DeVol 1999, p. 62). This represents a trend of continued concentration and strength. In 1990 almost one-third of U.S. employment in the film industry was located in Los Angeles, generating almost half of the income in this industry (Finney 1998). As mentioned already, the digitization process of all forms of information and their following fusion, have accentuated the notion that filmed entertainment, notably television, constitutes information. This form of information has become most important and influential. Thus, on an average day, an average American spends 217 minutes watching television, compared to only 123 minutes of listening to the radio, and much less time reading newspapers (29 minutes), or magazines (17 minutes) (*TV Basics* 2000).

Visual electronic entertainment may be conveniently divided into three major types: movies, television and the Internet, with a possible fourth category of multimedia games. All three peaked first in New York, which earlier enjoyed supremacy in the non-electronic visual entertainment, the theater (Christopherson and Storper 1986). However, once leadership was attained in motion picture production by Hollywood, it led to later leadership in television program production, and eventually to competition with New York on Internet supremacy. Theaters, however, remained concentrated in New York, being based on other technologies.

Compared to traditional FIRE information which could achieve concentration in New York already in the early 19th century, filmed entertainment was invented only in the late 19th century, concentrating first in New York, and peaking in Los Angeles as of the 1920s (Storper 1997). This move marked an early local specialization through geographical separation in information production. It reflected superior climate and landscape conditions in Los Angeles, as well as entrepreneurship of specific Jewish and other immigrants concentrated there (Abu-Lughod 1999).

In the mid-1940s the Los Angeles filming industry owned much of the American cinema industry, but this was followed by a decline in the industry in the late 1940s–early 1950s, mainly due to the competition with the new television industry, by then concentrated in New York. As of the late 1960s–early 1970s this new industry concentrated too in Los Angeles, except for news production which has remained mostly in New York (Storper 1997, Christopherson and Storper, 1986).

As of the 1990s, the major studios as well as numerous smaller ones, concentrated in the Los Angeles area (Storper and Christopherson 1987). However, their role has changed from production, which moved to other

PRODUCTION

locations throughout the U.S., to entrepreneurship and mobilization of venture capital (Storper 1997). Another trend has been the growth in the number of small companies providing specialized services, such as set production, model making, etc. (Kotkin 2000, p. 136). In this transition, the filmed entertainment industry became similar to other parts of the information economy, notably the Internet and start-up high-tech industries, in which initiation and capital raising is performed in major centers, such as New York and San Francisco, but production may take place elsewhere. Similar processes were identified for the magazine publishing industry, as we noted earlier. Thus information types not only fused with each other through digitization processes, but their economic-geographical organization became similar. More generally, information *content production* is concentrated in just a handful of cities, whereas information *(container) manufacturing* (books, magazines, cassettes, etc.) may be spread more widely.

Like the two previously invented forms of electronic entertainment, motion pictures and television, Los Angeles reached leadership in the Internet in a *second strike*. Comparing the 1994 ranking of *central cities* by the number of commercial domains with those of 1997 and 1999, New York was followed in 1994 by San Jose and San Francisco, with Los Angeles ranking 15! Three years later, in 1997, San Francisco was second, followed by Los Angeles, and Chicago (Moss and Townsend 1998), and in 1999 Los Angeles turned second, followed by San Francisco (Zook 1999, Tables 4.3 and 4.4). This may attest to the initial superior role of finances and high-tech over the media in bringing about a local specialization in Internet information production. The commercialization of the Internet and its contents, dealing mainly with entertainment, have given Los Angeles, with its media experience, a relative advantage.

Hollywood in Los Angeles, like Manhattan in New York, has become the locus for the filmed information industry. Its role changed from a manufacturing area to a kind of *downtown*, with premieres taking place there, and services for the industry located in it (Christopherson and Storper 1986). Not surprisingly, Hollywood (jointly with Santa Monica) has become the center for the Los Angeles Internet industry, enjoying its cumulative advantage (Zook 1998).

Another area in which Los Angeles at large and Hollywood in particular have enjoyed their cumulative advantage in audiovisual entertainment, is the multimedia industry (Scott 1995). In 1995 one-half of the U.S. multimedia businesses were reported to be located in California, divided equally between the San Francisco and Los Angeles urban areas, with the Hollywood-Santa

Monica area serving as the dominant concentration. These heavily computer-based industries also developed strong bonds with major rather traditional book publishers in New York (Scott 1995).

In the consumption of information, the pattern found for New York is repeated, namely that Los Angeles does not lead in the consumption of information types as it leads in production. The television as well as cable television markets of Los Angeles are second to those of New York (*Nielsen Media Research* 2000), which is quite obvious, given the difference in population size, and the 98% penetration rate of television in the U.S. In Internet penetration, Los Angeles ranked fifth in 1999, tenth in 2000, and lower in 2001 (see Kellerman 2000a; Table 8.1). The production of television and VCR sets is mostly located outside of the U.S., with only 26% of those products sold in the U.S. actually produced or assembled there (CEMA 1999).

4.6   IT R&D and Information Production

Comparing the conceptual and empirical parts of Chapters 3 and 4, notably Table 3.5 (U.S. city rankings by technology measures) with Table 4.4 (domains in U.S. metropolitan areas by percentage of national total), may reveal some interesting relationships between the R&D of information technology and the production of information. All the nine cities leading in Internet information production or packaging appear also on the various lists of innovation and technology measures. More striking is the correspondence between the leading cities in both areas, namely New York, Los Angeles, San Francisco, followed by Boston, Chicago, etc. (see also Table 8.1). If we take into account that another and non-American leading information city, London, is also a major technology center, then the comparison is even more widely begging.

One level of relationship seems most basic, namely client and server relations. The high-tech industry develops the major tools for the Internet information industry, whether it be hardware, software, or telecommunications. Furthermore, the high-tech industry may respond much better and faster to the needs of the information industry if the two are concentrated within the same cities. The importance of shared tacit knowledge, as well as informal contacts, has been accentuated for both sectors separately, and these may be of importance also for exchanges between the two industries. As Lundvall and Maskell (2000, p. 360) noted: 'innovations reflect a process where feedbacks from the market and knowledge inputs from users interact with knowledge creation and entrepreneurial initiatives on the supply side'. Another added dimension which we shall further see in Chapter 8, is that the high-tech information technology is a major user of the Internet and its innovations, something which may strengthen the bonds between the two sectors, at the local scene. This type of relationship is most evident in the San Francisco/San Jose area, in which the high-tech industry has become Internet oriented, and well geared with the leading information industry in the metropolitan area (see Saxenian 1994, pp. 117–118).

There is, however, another level in the relationships between the two sectors. Both have developed along similar organizational and cultural lines. They both require the investment of venture capital, notably in start-up enterprises. Both 'are dominated by the language of individual achievement' (Saxenian 1994, p. 164), channeled many times through small companies, and coupled with dynamism and flexibility, qualities which Saxenian (1994) found to be dominant in the Silicon Valley, and much less so in the Boston area. This may help to explain the superiority of San Francisco over Boston in the contemporary information economy, despite the Bostonian traditional specialization in book publishing. The close cultural and production relations between the two industries may further explain the rather large number of domain names in cities which are ranked lower in population size. In other words, local firms in high-tech cities may tend more to go on the Internet, either because of the availability of local expertise, or because of the business atmosphere in high-tech cities. The local Internet information industry enjoys a strong local business basis in addition to the packaging of information produced elsewhere.

The close relationship between the two industries is not a geographical and economic must for either of the two. A high-tech region may specialize in components that do not have direct or immediate implications for the Internet industry, and websites, on the other hand, may develop with the help of experts in the Internet only. What seems crucial, however, is that *leadership* in the information industry requires the geographical proximity of the high-tech information technology industry, for direct and indirect relations, facilitating constant innovations and pioneering in the computer-aided production and packaging of information.

4.7 Conclusion

The model presented in Figure 4.1 detailed several partners in the production, packaging and transmission of Web information, possibly located in different cities. However, the geographical convergence among the various phases from R&D, through production to consumption, is high, as will be shown in the following chapters on transmission (Chapter 6) and consumption (Chapter 8). The Internet has not constituted a revolutionary new technology, not based on previous technologies and knowledge. It was rather based on previously existing knowledge, technological as well as artistic, and it permitted the fusion of previously existing media. As such, its development and dissemination has flourished in cities with relevant cumulative advantage.

Zook (1998) raised the possibility of future diffusion of Internet production along the urban hierarchy as the industry matures, following product cycle theory. However, one may identify some similarities between financial centrality and Internet production centrality and in various aspects: the very existence of leading centers or the first mover advantage; the global location of customers; the absolute dependence on telecommunications for the transfer of capital and information respectively; and the accumulation of expertise in these centers. Thus, it might well be that the location of the Internet industry will disperse, as has really been the case in recent years, but on the other hand, leading centers in both information technology and information production will continue to pioneer new technologies and approaches, as well as dealing with the heavier and more complex websites, even if there is no significant association between the contents of Internet websites and the more veteran specializations of the cities, such as finances.

The portrayal of the two leading centers of production on the global information space, New York and Los Angeles, may lead to conclusions in three spheres: geographical–structural; regional; and processes leading to locational peaking. From a geographical–structural perspective, information industries which yield electronic products, such as capital, FIRE, and the Internet are concentrated in major centers through all their phases of production, processing and transmission. However, information industries which, while being computer-aided, yield material products such as books, magazines, disks and cassettes, may locate their manufacturing facilities elsewhere, so that leaders in the information economy serve as command and control centers, with entrepreneurship, capital mobilization, information production, and decision making concentrated in them. The geographical patterns of information consumption, as well as those of manufacturing of information hardware and software, may be rather different.

PRODUCTION

It is for very small areas within leading information cities, such as Manhattan in New York and Hollywood in Los Angeles, to serve as global
concentrations of information production, enjoying rather long cumulative
advantage. It remains to be seen in the next chapter whether the specific
long economic specializations of leading cities in the information economy
have had an imprint on the content of the Web.

5

'Web sites are a form of geography – geography of the screen'.

(Dodge 2001b, p. 173).

As mentioned already, the Internet consists of two interrelated information bodies. The first constitutes the electronic *transmission* of information at large, or of communicative materials, and the second is an information *system*, the Web. Information transmissions may be either e-mails, often in the form of flexible, unstructured business or personal communications, or they may constitute structured data files. One sort of data file is transfer of capital. On the other hand the Web, by its very nature, consists solely of structured information files organized through websites.

The following discussions will focus on several geographical aspects of this information transmission and system, notably the latter. First, the content of the Web will be examined by subjects, and a possible relationship between these subjects and the specializations of the leading cities in the production of information will be explored. Then, attention will focus on one form of transmission and its geographical ramifications, the capital industry. Third, e-commerce will be highlighted, e-commerce constituting a structured form of information through websites, but creating also significant Information traffic through ordering and enquiring by customers. Finally, the language of Internet use and operation will be discussed from a geographical perspective.

5.1 Content Demand and Location

Some 36 million websites were reported by the *Oracle* company to exist on the Internet at the beginning of 2002 (Doron 2002). A possible connection between Internet content and location of production was raised by Zook

(1998): 'one would expect different types of Internet content based on what activities were already present' (p. 20, see also Zook 2000). It was observed in June 1997 that 40 of the most accessed 100 websites were located in California, a finding which fits the geographical pattern of Internet production (OECD 1997). What can one guess about the contents of websites? The concentration of Internet information production in cities with particular related specializations could have been expressed in sizeable numbers of websites devoted to financial information, high-tech topics and multimedia presentations. In addition, the leadership of these cities in R&D may suggest high percentages of websites devoted to academic-educational materials (Kellerman 2000a). Zook (2000) noted in this regard that 'it appears that there is a stronger connection between Internet content and information-intensive industries than between the Internet and the industries providing the computer and telecommunications technology necessary for the Internet to operate' (pp. 411–412).

In order to examine a possible fit between location and contents, three dimensions may be looked at: the topical distribution of the Web; the most frequent terms sought for through search engines; and the most visited sites. If the high share of domains devoted to commercial sites is applied also to the distribution of pages, then a very high majority of the Web is commercial. A survey conducted by the Israeli *Rivlin* company back in 1999 revealed the distribution of Web pages at large (both commercial and others) by leading subjects (*Ma'ariv*, 1999) (Table 5.1). The richness of the Web was found to be extremely wide, since the twelve most popular topics accounted for only one-third of the pages, and since sex, the most popular subject, reached only 5% of the total number of pages. Finances, the most important telecommunications-associated activity and the most striking one in the major Internet production centers, is not included among the 12 leading subjects. Education, however, ranks second with some 4.5%. Computers and advertising, the two closest topics to high-tech and multimedia respectively, close the list, with each having just about one quarter of the number of pages devoted to sex.

Between August 1998 and August 1999 the fastest growing category in terms of site visiting was shopping (35%), followed by travel (29%), with sex growing the least (4.6%) (*Media Matrix* 1999). Leading cities in the number of website pages devoted to them, though relating more to travel and tourism than to business, included in 1999 only one of the three leading centers of Internet production: New York, and only as third (4.58M pages), following London (6.1M pages) and Paris (4.6M pages).

Table 5.1 The most popular subjects on the web, 1999

Subject	No. of pages (in millions)	%
Sex	42.6	5.0
Education	36.8	4.5
Music	34.7	4.2
Management	32	4.0
Health	26.7	3.3
Travel	20.2	2.5
Manufacturing	20	2.4
Entertainment	15	1.8
Shopping	12	1.5
Food	12	1.5
Computers	11.8	1.4
Advertising	11	1.3
Total	**274.8**	**33.4**

Source: Ma'ariv (1999).

Table 5.2 Subjects of non-commercial websites, 1999

Subject	% of webservers
Scientific/educational	6.0
Health	2.8
Personal	2.3
Societies	1.8
Pornography	1.5
Community	1.3
Government	1.2
Religion	0.8
Total	**17.7**

Source: Based on Figure 1 in Lawrence and Giles (1999).

Another study focused on the contents of circa 15–18% of the non-commercial Web information (Lawrence and Giles 1999) (Table 5.2). Obviously, most of the subjects in these websites are different from those of the Web at large, which consists of an overwhelming amount of commercial information: they are less entertaining, and more of an interpersonal and 'serious' nature. However, three subjects appear on both lists: education, health and sex. Educational/scientific materials lead the non-commercial list with 6%, compared to just 4.5% on the general list. This difference may be attributed to different measurement methods: the *Rivlin* rather general survey used search engine information inflated for the whole system, whereas the Lawrence and Giles study examined the contents of a sample of 2500 servers. In general, sex and scientific materials competed as leading subjects on the Internet, attesting to the wide spectrum of information on the system. Obviously, most of the materials on sex are commercial, which is true also for health, though at lower proportions.

Most common terms searched for on major search engines have an even wider distribution. Thus, the 25 most searched terms on *Alta Vista* comprised only 1.5% of all queries (see Paltridge 1999). Websites devoted to the provision of information on the most frequently searched terms are mostly censored to exclude terms of sexual content. However, a leaked list of the most searched terms on *Yahoo!* in October 1996 put the word 'sex' in the first place with 1 553 420 searches, followed by the word 'chat', with just one quarter of the searches for 'sex', 414 320 (Eyescream 1997). In this list some 60% of the 200 leading terms had a sexual connotation! The search engine *Lycos* provides weekly statistics on the 50 most searched terms. Table 5.3 presents a comparison between the most searched terms in October 1999 and October 2001. The 1999 list is comprised mainly of entertainment and music terms, whereas the 2001 list includes also many items relating to the terror and war events of Fall 2001. Interestingly, *Lycos* reported on several of the entertainment/music items of early October 2001 as being on the list for the 112th week!

Table 5.3 Most searched terms on Lycos

October 1999	October 2001
Pokemon	World Trade Center
Halloween	Halloween
Dragonball Z	Osama Bin Laden
Britney Spears	Dragonball
WWF	American flag
Pamela Anderson	Costumes
NFL	Morpheus
Beanie Babies	Nostradamus
Backstreet Boys	NFL
Poetry	Afghanistan

Source: *Lycos* (1999, 2001).

Table 5.4 Most accessed websites

August 1999	September 2001
Yahoo!	Microsoft sites
AOL	AOL Time-Warner networks
MSN	Yahoo!
Geocities	X10.com
Netscape	Excite network
Go	Lycos sites
Microsoft	About/Primedia
Lycos	Terra Lycos
Excite	e-Bay
Hotmail	CNET network

Source: Media Matrix (1999, 2001).

Most accessed websites are reported, for example, by *Media Matrix* (1999, 2001). Table 5.4 provides a comparison in this regard between August 1999 and September 2001. In both years the leading sites have been search engines and other 'general purpose' sites or portals. Thus, in both years, only in the 12–13th rank did a more specific site appear: *Amazon.com*. It was estimated for October 2000 that 86% of all top websites were American, mostly in shopping and finances (Zook 2001c).

The contents of the Web and the demand for information attest to a most diversified system geared foremost to leisure activities led by sex, inter-personal communications, travel, music, and shopping. Secondary to these activities are educational-academic materials. Finance, the leading economic activity in the centers of Internet information production, is not among the leading types of produced or consumed Web information, even though it may lead in other forms of electronic communications, such as data and telephony. Computers and high-tech, constituting other leading specializations in Internet information production, are also not in the lead as far as Web information contents, whereas the media are represented, at least in the form of music sites. In other words, in terms of Web contents, the Internet industry has been driven by a wide and popular demand for information, independent of the economic specializations of the major information production centers. This independence has emerged despite the young age of the Web as an infor-mation system. The system rather depends on free and mainly commercial marketing of information, services and products to global markets.

The leading Internet sex industry constitutes an exception in this regard, namely that previous specialization in content production has been followed by website production, as was shown by Zook's (2001c) analysis. It was variously estimated that 25–39% of Internet users surfed sex websites. The sex information industry is typified by sensitivity to local regulations and cultural sensitivities and is thus a volatile one. It further requires high bandwidth and strong hosting servers, given the extensive graphics involved. As with other types of content, the U.S. leads in adult content, with 69.8%, followed by Canada (9.0%), Netherlands (2.9%), Australia (2.8%), and U.K. (2.2%). Within the U.S., the dominance of Los Angeles is striking (Table 5.5), with 23.2% of U.S. contents production, or one-third of global production! This is clearly related to the region's predominance in contents production through an earlier technology of video production. The industry concentrated in the San Fernando Valley as of the mid-1980s. O'Toole (1998) relates this concentration to nearby Hollywood, and its supply of filming professionals, models and equipment, rather than a feeling of safety in a large concentration. New York and San Francisco followed in video production, as well as in online content production, though in much lower percentages, and the industry at large seems to decentralize with the move from video to Internet production.

Another type of content which is considered illegal in various countries, states, or cities, is virtual gambling, the casinos of which are located in areas where this activity is permitted, whereas gamblers may be located elsewhere (Wilson 2002). The distribution of casino domains was led in 2001 by the U.S. with 278 sites, or 23.4% of the 1186 such sites worldwide, followed by Canada with 253 (21.3%), and Antigua with 110 (9.3%). North America and the Caribbean dominated the scene, but the urban distribution in the U.S. is widely distributed compared to the Internet industry at large (Table 5.6). Canada turns out to share leadership in this Internet application, and Montreal

Table 5.5   The location of video adult content providers vs. online content providers, 2000

MSA/CMSA	Video content providers	Online content providers
Los Angeles CMSA	56.0	23.2
New York CMSA	9.0	7.1
San Francisco Bay CMSA	5.3	2.4
San Diego	2.9	4.5
Seattle CMSA	2.7	4.7
Las Vegas	2.5	4.2
Total	**78.4**	**46.1**

Source: Zook (2001c).

Table 5.6 The location of virtual casino domains, 2001

City	No. of domains	% of national total	% of world total
Montreal, Canada	108	42.7	9.1
San Jose, Costa Rica	84	93.3	7.1
Basseterre, St. Kitts and Nevis	47	78.3	4.0
Panama City, Panama	39	100.0	3.3
Toronto, Canada	37	14.6	3.1
Winnipeg, Canada	32	12.6	2.7
Miami-Ft. Lauderdale, U.S.	29	10.4	2.4
Los Angeles, U.S.	28	10.1	2.4
London, U.K.	25	34.7	2.1

Source: Wilson (2002).

is the globally leading city with 9.1% of world sites and 42.7% of Canadian sites located there. In smaller Caribbean nations the urban concentration is very high, given the smaller urban systems in these countries, and the more enhanced telecommunications infrastructure available in major cities.

5.2 Capital as Information

5.2.1 Capital, space and information

International flows of capital on a global scale date back several centuries. In the nineteenth century these flows served and even constituted processes of colonization and empire building. In the second part of the twentieth century they flourished and intensified as a result of national policies, technological development and market growth. The interrelationship between capital, information, and space is of much interest. 'The great power of capital has always been its ability to choose, to decide where to locate' (Mulgan 1989, p. 19). This ability has been enhanced by innovative information and communications technologies. Furthermore, information networks are required for capital to be valid over wider space (Dodd 1994, p. 159). Such an expanded validity extends the ability of money to choose investment locations, which, on their part, may call for additional information and capital flows.

By its very nature, capital is both a resource and a product. As mentioned earlier, capital has also constituted the heaviest user of telecommunications. Capital has further been the only material thing, though a symbolic one, comprised of paper notes and coins, that has been transformed into electronic bits, or information: 'nowadays money is essentially information' (Thrift 1995, p. 27). As such, capital may be highlighted from three inter-related perspectives: first, its *flows*, notably at the global scale, facilitated by telecommunications; second, its *places of management*, which coordinate and control increased global capital flows, and which are simultaneously reinforced by them. The management of capital has been channeled recently also through the Internet as a virtual place. Third, capital management is embedded in worldwide *urban expressions*, embodying global capital invest-ment. These amount to physical expressions of the global (capital and its derivatives) in the local (urban landscape).

Capital from various cities and countries is accumulated into global capital through international banks, funds, and corporations, a process which has been permitted through the collapse of spatial national barriers for capital flow and assembly. This global capital is then invested in specific locations. Whereas the development of global capital may involve the production of space or place in a rather limited number of global capital centers, such as New York, London, or Tokyo (see e.g. Sassen 1991, Kellerman 1993a), the flow of global capital into local communities may contribute to the production of place in a potentially large number of cities.

5.2.2 Flows

International financial markets have developed since the late nineteenth century, but almost disappeared between the early 1930s and late 1950s, the period of economic crisis and war. After the late 1950s, governmental policies reduced regulatory barriers on capital movement. Coupled with mar-ket forces and improved international telecommunications and information technologies, they have facilitated extensive capital flows on a global scale (Helleiner 1994). These flows have eroded the meaning of political bound-aries as a means for controlling money flows by nation-states (Kobrin 1997, Warf and Purcell 2001). Capital flows have been assumed to regulate national economies and business activity (Amin and Thrift, 1994). A different view was Gordon's (1988), who argued at the time that the global economy was less open and multinational corporations less powerful than normally assumed. He, as well as Cox (1992, 1993), further argued that *productive capital*, as opposed to *financial capital*, was not *hypermobile*. However, one of the char-acteristics of the contemporary mobility of capital and the sophistication of investment vehicles is that the differences between these two types of capital are getting blurred. The shift of power from national regulating bodies to financial institutions has marked a shift of power from the national scale to the global and local scales (Swyngedouw 1992a).

This change permits viewing developed countries as functioning globally on one barrier-free, almost isotropic, plain. 'The globalization trend in finance has somehow been beyond politics' (Helleiner 1994, p. 2). In this, the global flow-space of capital behaves similarly to that of computerized information via the Internet, whereas the equivalent international movements of commodities and people are still nationally regulated (see Appadurai 1990).

Global electronic flows of capital date back to the 1970s, institutionalized in the mid-1970s with the internationalization of *Visa* and *Mastercard*, and even more so with the establishment in 1977 of *SWIFT* (Society of Worldwide Interbank Financial Telecommunications), as an international interbank electronic funds transfer system (see Kellerman 1993a, p. 109). Detailed and reliable data on the geographical origins and destinations of global capital flows actually do not exist. It was estimated that around $1.3 trillion flow around the globe daily, out of total global financial assets of around $66.4 trillion in 1993/94 (see Laulajainen 1998, p. 28). Thus, about 2% of global assets flow daily. The World Bank, as well as the International Monetary Fund (IMF) have refrained from the publication of geographical breakdowns of the daily or aggregate annual global flows of capital. Partial estimates for the 1980s and early 1990s were presented by Laulajainen (1998, pp. 33–34). Aggregated annual data of in-flows and out-flows of equities for 1988 for four global financial cores permitted the calculation of the following net balances (in $billion) per core area: the U.S. gained 134, Japan gained 66, the U.K. lost 116, and the EU lost 84 (see Laulajainen 1998, p. 34).

Sassen (1994, pp. 10–18) used United Nations data on foreign direct investments (FDI), (that is foreign investments in new or existing firms), in order to study geographical trends in foreign investments. In 1990 FDI reached $203 billion, out of which some $172 billion, or 84.7% were invested in the three global cores, namely the U.S., Western Europe, and Japan. From the 1950s onwards, this geographical concentration tended to increase. Four countries, U.S., U.K., France and Germany, received half of the world FDI in the 1980s, with Western Europe being the leading recipient world region. Five countries, U.S., U.K., France, Germany, and Japan, accounted for 70% of the exported capital. In other words, much of the FDI capital circulated among the four most developed national economies, whereas the receiving share of developing countries decreased from 26% to 17% between the early and late 1980s. The late 1990s marked a notable growth in FDIs in South

America, but again concentrated in specific countries: Brazil, Mexico, and Argentina (Sassen 2001b).

The growth in FDI has been very impressive: it nearly tripled between 1984 and 1987, and it grew by 20% annually 1988–1989. The fast growth in global capital flows has been attributed to the emergence of the Pacific Rim, notably Japan, as a capital exporting region, the emergence of multinational corporations, and the growth of the service economy.

In contrast to the flows of FDI which tend to 'land' in the major national economies, fast moving *furtive money*, that is capital originating in the major national economies but seeking to avoid regulatory attention and/or taxes, flows to various mini-states which serve as offshore tax havens (e.g. Hong Kong, Singapore, Liechtenstein, Gibraltar, the Bahamas) (Roberts 1994). For example, the Cayman Islands served in the early 1990s as the home of $250 billion in bank liabilities. The growth in offshore banking has been attributed mainly to the growth in *Euromarkets*, and thus to the flows of FDI. The offshore mini-states are dependent on the time zone and related stock market operating hours in the major markets geographically adjacent to their location, notably those of New York, London, and Tokyo.

5.2.3 Management

Capital requires and produces places for the management and coordination of its flows over space (see Sassen 2001b). This geographical requirement for concrete locations to facilitate flows, is making global space become both *reflected* and *anchored* in local space as place:

> *Capital is an inexorable process diffusive in space which also fixates itself as a thing in space and so begets a built environment. The fixity nature (the thing quality) of the geographical landscape is necessary to permit the flow and diffusive nature of capital; and vice versa. Capital fixity must, of necessity, take place somewhere, and hence place can be taken as a specific form emergent from an apparent stopping of, or as one specific moment in, the dynamics of capitalist social space... The* production of space *is thus the process as well as the outcome of the process (i.e. the produced social space); it is the totality of the 'flow' and 'thing' qualities of capitalist material geographical landscape.*
> (Merrifield 1993, p. 521, following Lefebvre 1991, pp. 86–92).

The emergence of a world financial center (other than offshore) is a five-phase process, moving from the service of a local area, through the service of a region, a nation, neighboring countries, to global service (Reed 1981). Thus, global financial centers enjoy a cumulative advantage in their prolonged development. In addition to its historical development, a world financial center must possess information and communications centrality, as well as its being a major capital importer and exporter (Laulajainen 1998, pp. 253–255; Grote *et al.* 2002). Though about one-half of world equities are spread in about 100 cities worldwide (by the location of the fund), trade in them tends to be much more concentrated in a handful of stock markets, led by New York, London, Tokyo, Paris, Zurich, Frankfurt and Hong Kong. This stems from the need to assure high levels of liquidity, which on its part leads to the development of local expertise and competition (Laulajainen 1998, pp. 257–258). In the FX (foreign exchange) area there could potentially evolve a complete centralized trading in one node, but the current hub and spoke pattern has been preferred, since it permits conventional working hours in each part of the world, as well as sensitivity to domestic and regional business and political events (Langdale 2001).

Another important dimension of global centers for capital management is the need for *social connectivity* coupled with *financial technology* (Sassen 2001b, Grote *et al.* 2002). As Thrift (1994a, p. 334) noted: 'the need for information, for the expertise that allows that information to be interpreted and for the social contacts that generate trust, information, interpretive schemes – and business'. Whereas data, information, and codified knowledge may be transmitted electronically, it is for financial tacit knowledge, shared through face-to-face meetings of the kind made possible in major centers, where centrality becomes crucial. These contacts are coupled with the expertise of accompanying producer services located in major urban financial centers. Grote *et al.* (2002) argue further for the importance of *cultural proximity* among actors in financial markets, in addition to the *spatial proximity*. These two proximities may, though, be partially substituted by an *organizational proximity*, namely workers of the same company operating in different locations.

The same improved telecommunications technologies which permit global and national capital flows may be used for intracity flows as well, and thus they may facilitate expanded and more efficient flows and a restructuring of metropolitan spatial organization. Growth in activities related to global capital normally takes place in CBDs, through major domestic and foreign banks, stock markets, consulting and investment firms, and the like. Thus, domestic businesses have to compete on rising rents. Since these domestic companies too require proximities among themselves they contribute to the emergence of suburban *mini-cities* or *edge cities* (see e.g. Garreau 1991). Furthermore, rising costs of downtown space, coupled with the availability of information technologies and networks have brought about the emergence of *back offices*, so that divisions such as manpower, accounting, etc. of companies headquartered in CBDs move to suburban locations (see e.g. Kellerman 1993a).

The introduction of the Internet has permitted a change in the management of financial business. Whereas telecommunications and *SWIFT* have permitted inter and intrabank business, the Internet has facilitated direct and real-time electronic connections between customers and financial institutions, as well as trading in areas such as FX (Langdale 2001). These types of interaction have been further developed with the introduction of secure Web-based environments (Grote *et al.* 2002). It was estimated that the cost of an Internet banking transaction is merely 1% of the same performed physically by a teller (Leinbach 2001, p. 21). Internet technologies may permit, in upcoming years, the introduction of global services for the complete management of customers' assets, without regard to location of either customers, investments or management services. Once again, however, companies located in major financial centers will be able to provide a more professional and updated service. The introduction of supranational currencies such as the Euro may assist the development of such services, which imply further investments in R&D as well as production of information and communications technologies.

5.2.4   Urban expressions

A constant tension exists between the globalization and localization of capital: 'While capital expands over absolute space, extending its control over space, it simultaneously has to engage in a struggle over space… The wider capital's control over space, the more important the place-specific conditions for accumulation become' (Swyngedouw 1992b, p. 60). The material fusion between the global and the local takes place, therefore, in the local or urban arena. Harvey (1989, pp. 293–296) pointed to the importance of space when global spatial barriers collapse. 'As spatial barriers diminish so we become much more sensitized to what the world's spaces contain' (p. 294; see also Dodge and Kitchin 2001, p. 35). This view indicates a transformation in the meaning of space from barrier, on the global scale, to container and place, on the local one. Harvey (1989) accentuated the constitution and meaning of space as relative location. 'Heightened competition under conditions of crisis has coerced capitalists into paying much closer attention to relative locational advantages' (pp. 293–294), which may lead to the production of places:

> *The qualities of place stand thereby to be emphasized in the midst of the increasing abstractions of space. The active production of places with special qualities becomes an important stake in spatial competition between localities, cities, regions, and nations... Heightened inter-place competition should lead to the production of more variegated spaces within the increasing homogeneity of international exchange... We thus approach the central paradox: the less important the spatial barriers, the greater the sensitivity of capital to the variations of place within space, and the greater the incentive for places to be differentiated in ways attractive to capital.*
> *(Harvey 1989, pp. 295–296).*

Harvey, then, views the relationship between global/national and local spaces as a one-way *causal* relationship leading from the global to the local, so that the homogenization of the global should lead to the *specialization* of the local, through the specific advantages of any given location. Furthermore, capital can also, by its very nature, *create* locational advantages and specialization. Swyngedouw (1992a) noted Marx's space/technology nexus, in which 'the advantages of "better locations" can be offset by capital investments in "inferior locations", or location and investment can actually reinforce each other... Marx identifies here also the basis of the trade-off between space and technology, a trade-off which makes location and technology to a certain extent interchangeable' (p. 424).

Specializations of urban economies within national or global contexts are not novel as far as shipping, tourism and industrial production are concerned. However, the introduction of direct and reasonably priced international telecommunications, and the resulting increased global flows of capital have brought about global specializations in the management and transfer of capital, as well as an emergence of a global urban hierarchy consisting of domestic, world, regional, and global cities (see Reed 1981, Sassen 1991, Kellerman 1993a). The triad of New York, London and Tokyo leads the global financial economy, with Tokyo leading in the production of capital through the late 1990s, London specializing in the management of global capital through international banking, and New York leading in investment markets (see Kellerman 1993a, pp. 102–103). However, besides this triad,

Luxembourg, Singapore and Hong Kong lead in FX trade (Laulajainen 1998, pp. 107–109).

Global flows of capital and their anchoring in cities imply construction activity. Directly related construction activity includes banks, headquarters and domestic offices of multinational corporations and business hotels. These buildings are largely located in central business districts or edge cities, which together with other office concentrations serve as 'containers' for the controlling of investments of global capital. The international nature of these buildings, their foreign financing, and the global flow of architects and designers contribute to a homogenization of urban office landscape, giving an impression of 'the world as a single place' (King 1990). Such similarities in architectural styles have continued to sustain even under postmodernism (Castells 2000, p. 448). Similar processes have been identified also for high-tech parks, imitating Silicon Valley, in the interior design of buildings, as well as in their architectural style (Graham and Marvin 2001, pp. 333–337). Such processes may be amplified by extensive foreign buying of real estate in downtown areas (e.g. Los Angeles and London), sometimes representing speculation and fears of devaluation (Appadurai 1990, Swyngedouw 1992a).

Increased local investments, economic activity, and geographical expansion, may yield capital flows out of, as well as into, localities. Harvey (1985, see also Gregory 1994, p. 93) commented on the spatial barriers which may be imposed on these continuous two-way flows:

> *The produced geographical landscape constituted by fixed and immobile capital is both the crowning glory of past capitalist development and a prison that inhibits the further progress of accumulation precisely because it creates spatial barriers where there were none before. The very production of this landscape, so vital to accumulation, is in the end antithetical to the tearing down of spatial barriers and the annihilation of space.*
>
> *(Harvey 1985, p. 43).*

However, it may well happen that such 'old' and spatially 'imprisoned' investments actually attract additional flows of global capital. This may be the case with investments in resorts and tourist attractions, representing the interrelations between global flows of capital and people.

5.3   E-Commerce and Location

The *Internet economy* has been defined as 'economic developments involving the use of the Internet as an important business driver' (Barua *et al.* 1999, p. 3). This rather wide definition consists mostly of electronic commerce (e-commerce), but also of the distribution of information and informative interactions between suppliers and customers, which we have noticed regarding information types (Chapter 1). The Internet economy has grown fast, notably in the U.S., prior to the 2001 slow down. Thus, in 1998, just four years after its early emergence, the U.S. Internet economy was estimated at 301 billion dollars, more than energy, automobiles and telecommunications (Barua *et al.* 1999). Despite its high-tech image, employment in the U.S. Internet economy in the first half of 2000 consisted of 28% information technology workers, 33% sales and marketing workers, and only 9.6% workers of *dot-com* companies (University of Texas, 2001). The following discussion will highlight e-commerce within the wider context of the Internet economy, focusing on its nature, structure, geography, and major actors.

5.3.1   Nature

Electronic commerce initially began in the 1970s, with the French *Minitel* system, based on computer terminals and the telephone system, and it boomed with the introduction of the Internet in the mid-1990s. E-commerce has been variously defined within the two contexts in which it is anchored: economy and information. One economic definition puts within e-commerce 'any form of economic activity conducted *via* electronic connections' (Wigand 1997, p. 2). This spectrum was defined from the perspective of information technology as: 'the automation of commercial transactions using computer and communications technologies' (Westland and Clark 1999, p. 1). Yet another economic definition includes within e-commerce banking, as well as commercial activities through television (Palmer 1997, p. 75).

Defining e-commerce from an information perspective has received both narrower and wider definitions. On the one hand, e-commerce may be viewed as *based* on the electronic transfer of information: 'commercial activities conducted through an exchange of information generated, stored, or communicated by electronic, optical, or analogous means, including electronic data interchange (EDI), electronic mail (e-mail), and so forth' (Hill 1997,

127

p. 33). On the other hand, however, the transfer of information electronically per se may be considered e-commerce: 'We consider e-commerce as the collective of numerous individual, not always commercial, electronic strands of information that move at local, regional, and global scales between and among individuals, universities, corporations, non governmental organizations (NGOs), and states for a variety of purposes' (Leinbach and Brunn 2001, p. xii). In the next section we shall examine e-commerce along the lines proposed by the Center for Research in Electronic Commerce, University of Texas at Austin (2001, Barua *et al.* 1999), namely that e-commerce consists of the commercial provision of network and WWW infrastructures, which may serve non-commercial information activities, as well as the sale of services and products through the Internet.

E-commerce may be typified by several major attributes: speed, accessibility or ubiquity, globalization, interactivity, information and intelligence (Leinbach and Brunn 2001, Kenney and Curry 2001). Several of these attributes are common to other forms of commerce as well, but are rather accentuated when the Internet is in use. This applies mainly to speed, accessibility, and interactivity, which are relevant in the physical world of both wholesale and retail trade, but receive a special importance when virtual-electronic systems permit faster and wider connections and responses. In a different way, though, information too is relevant to both physical and electronic commerce, since it is required for customers in any market activity (Wigand 1997). However, its wide and on-time availability in e-commerce has brought some commentators, as we have noted, to view all information exchanges over the Internet as e-commerce. Globalization is quite unique to e-commerce, permitting the creation of global market areas for many products and services. Another unique attribute is intelligence, whereby e-commerce is aided by sophisticated information tools, such as databases, search engines, and the like.

5.3.2 Structure

A common classification of e-commerce is by buyers into B2B *(business to business)* and B2C *(business to customers)*. Between the two, the larger share is of B2B, though B2C receives more attention, both on the Web, through advertising, as well as in research. In 1998, B2C comprised 15.7% of U.S. e-commerce, totaling $51 billion. The estimate of B2C for 2001 was just 9.4% of the total e-commerce, which increased in four years almost eleven times to $551 billion (NUA 2001). .

E-commerce, notably B2B among computer companies, has permitted a reorganization of company structure around networks. It permits keeping the scales of operation flexible, while providing for interactivity and flexible management, as well as the sustaining of brand names and B2C customization (Castells 2001). B2C is concentrated mainly in intangible products and services, such as travel, software, and entertainment (including music, gambling, and pornography) (Malecki 2000b). In fact, the 2001 slowdown hit grocery suppliers through the Internet. The 1998 breakdown and figures for U.S. spending online reveal some interesting patterns (Table 5.7). Travel was by far the most popular e-commerce application with over one quarter of the sales, followed by the partially intangible computer hardware and software products. These two industries comprised almost one-half of e-commerce sales. Information per se, or content, was the lowest category, but the larger family of information products (content, computer products, books, and entertainment) amounted to 44.8% of all sales! The sales of clothing, which enjoyed early starts of direct sales through catalogue and telephone sales, ranked rather low.

The Internet economy may be divided into four layers, the first two of which serve the Web as an information system at large, and on top of which two layers constitute e-commerce (Barua *et al.* 1999, University of Texas 2001) (Table 5.8). The first layer consists of products and services which help create an IP-based network infrastructure. The second layer includes products and services permitting the performance of business activities online. The third layer includes intermediaries which facilitate the meeting and interaction of buyers and sellers over the Internet, whereas the fourth

Table 5.7 U.S. consumer spending online, 1998 (in $million)

Product	Spending	%
Travel	1355	27.5
PC (hardware and software)	1085	22.0
Gifts and flowers	636	12.9
Entertainment	562	11.4
Books and others	518	10.5
Grocery foods	414	8.4
Clothing	316	6.4
Content	44	0.9
Total	**4930**	**100.0**

Source: NUA (2001).

129

CONTENTS

Table 5.8 E-commerce layers

Layer one: The Internet infrastructure layer
Internet backbone providers (e.g., Qwest, MCI Worldcom)
Internet service providers (e.g., Mindspring, AOL, Earthlink)
Networking hardware and software companies (e.g., Cisco, Lucent, 3Com)
PC and Server manufacturers (e.g., Dell, Compaq, HP)
Security vendors (e.g., Axent, Checkpoint, Network Associates)
Fiber optics makers (e.g., Corning)
Line acceleration hardware manufacturers (e.g., Ciena, Tellabs, Pairgain)

Layer two: The Internet applications layer
Internet consultants (e.g., USWeb/CKS, Scient, etc)
Internet commerce applications (e.g., Netscape, Microsoft, Sun, IBM)
Multimedia applications (e.g., RealNetworks, Macromedia)
Web development software (e.g., Adobe, NetObjects, Allaire, Vignette)
Search engine software (e.g., Inktomi, Verity)
Online training (e.g., Sylvan Prometric, Assymetrix)
Web-enabled databases (e.g., Oracle, IBM DB2, Microsoft SQL Server)

Layer three: The Internet intermediary layer
Market makers in vertical industries (e.g., VerticalNet, PCOrder)
Online travel agents (e.g., TravelWeb.com, 1Travel.com)
Online brokerages (e.g., E*Trade, Schwab.com, DLJDirect)
Content aggregators (e.g., Cnet, ZDnet, Broadcast.com)
Portals/Content providers (e.g., Yahoo, Excite, Geocities)
Internet ad brokers (e.g., Doubleclick, 24/7 Media)
Online advertising (e.g., Yahoo, ESPNSportszone)

Layer four: The Internet commerce layer
E-tailers (e.g., Amazon.com, eToys.com)
Manufacturers selling online (e.g., Cisco, Dell, IBM)
Fee/Subscription-based companies (e.g., thestreet.com, WSJ.com)
Airlines selling online tickets
Online entertainment and professional services

Source: Adapted from Barua *et al.* (1999).

layer constitutes the very sale of products and services to consumers or businesses over the Internet. Various companies operate at more than one layer. For instance, Microsoft and IBM are important players in Internet infrastructure and applications, as well as in Internet commerce. The four layers together were estimated to yield $830 million in 2000 in the U.S., with layers one and four, or infrastructure and e-commerce, each about twice in size the revenues of layers two and three, or application and intermediary services. In 2000, some 3125 *dot com* companies, mostly small ones, defined as layers three and four businesses, were counted by the survey conducted by the University of Texas (2001).

5.3.3 Geography

Coe and Yeung (2001) argued for the concentration of e-commerce in world cities due to their *first mover* advantage. Their study further counted three major geographical attributes of places specializing in e-commerce: the provision of relevant face-to-face contacts; the provision of logistics and distribution of products and services, including information technologies; and the existence of pertinent commercial initiatives and institutional support. Their analysis of Singapore accentuated the role of governmental support.

These attributes apply mainly to layer four of e-commerce. However, companies in all four layers seem to prefer locations in knowledge and information cities, or in global cities. Thus, it was reported for server farm companies, belonging to layer one, or infrastructure business, that they prefer highly secure building complexes located in global cities, which provide proximity to users, 'a factor of continuing importance in the location of heavily trafficked websites because of Internet congestion, bandwidth bottlenecks and the dominance of global telecoms capacity by major metropolitan regions' (Graham and Marvin 2001, p. 370). Another preferred location for server farms, as well as databases, are supposedly secure remote and peripheral areas, such as off-shore island states, like Bermuda and Anguilla, or even disused forts, such as the *Principality of Sealand* off the coast of England (Graham and Marvin 2001, p. 372). CBDs of global cities are preferred also by Internet backbone providers, such as *MCI Worldcom* (New York and London), or Internet service providers, (AOL in New York). On the other hand, PC and fiber optic producers would prefer technopoles, or industrial areas in knowledge producing cities, as we have noted already in our discussion of New York (Chapter 3).

Layer two, or application companies, show similar locational patterns. Of the Internet commerce application companies, *IBM* for example, is headquartered in suburban New York, and *Microsoft* is headquartered in Seattle. *Oracle*, producing web-enabled databases, is located in suburban San Francisco, and *Macromedia*, a major multimedia company, is located in San Francisco. Layer three companies, or Internet intermediary companies, were found to concentrate in downtown areas, attracted by the availability of a proper working force and business environment (Kotkin 2000, Gorman 2001). The top five American cities in which these companies concentrated were San Francisco, New York, Los Angeles, Boston and Chicago (Gorman 2001). Since most websites are commercial this concentration fits the list of cities leading in knowledge and information production (Tables 3.5 and 4.4).

Layer four companies, or Internet commerce companies, selling products online, require large storage, processing and distribution areas, located in the proximity of transportation hubs and networks. They may take over existing but declining such areas. However, since they require also the services of heavy telecommunications systems and Internet computer systems, they are once again located in warehouse areas in or in the vicinity of major cities, such as the American *Amazon.com* in Seattle, and the British one in Slough (Graham and Marvin 2001, p. 371).

The geographical difference among companies in the four layers of the Internet economy is, thus, more in intracity *siting* than in intercity *location*. Cities preferred by Internet business are global and information cities, but within cities, the production of hardware tends to concentrate in specialized industrial parks, server farms require special secure buildings in central areas, service providers and company headquarters also prefer downtowns, whereas distribution centers are located in warehouses, situated in more peripheral parts of metropolitan areas.

E-commerce over the Internet has flourished in the U.S. more than in other parts of the world, given the faster adoption of Internet connectivity, as well as a faster adoption of the Internet as business and communications means. In 1998, total consumer spending in the U.S. was estimated at $4930 million, whereas the equivalent figure for Europe was $298 million. Germany led with $160 million, or 53.7% of European e-commerce, followed by the U.K. with $60 million, or 20.1%. Other major European countries, such as France and Italy presented lower figures, 20 and 4 million dollars respectively (NUA 2001). The leadership of Germany and the U.K. was in line with their leadership in information production in Europe (though in an opposite order). Foreign sales by American companies online was estimated for 2000, between 20 and 25% of total sales (The University of Texas 2001). This rather low percentage, despite the global nature of e-commerce, may reflect lower home Internet connections in other parts of the world, slower adoption of e-commerce, and preference for domestic companies.

5.3.4 Vendors

E-commerce is not only concentrated in specific commercial branches and places of operation, but also in a handful of leading vendors. Some 75%

of U.S. e-commerce was estimated to be done through five websites: *Amazon.com*; *eBay*; *AOL*; *Yahoo!*; and *Buy.com*. (Button and Taylor 2001). Their location reflects once again the leading cities in the geography of knowledge and information production. *eBay* and *Yahoo!* are located in San Jose, *Amazon.com* in Seattle, AOL in New York, and *Buy.com* in Aliso Viejo, Orange County, California. It was estimated for the fourth quarter of 2000 that some 8 131 565 pageviews were made at the *eBay.com* site, followed by *Amazon.com* (3 199 398), *travelocity.com* (710 668), and *expedia.com* (663 647) (Alexa Research 2001), presenting once again two strikingly leading vendors. Pageviews do not reflect the number of shoppers, as only 1.6% of site visits result in purchases (Westland and Clark 1999, p. 4). *Amazon.com*, which operates two distribution centers in the U.K. and Germany, has its supposedly British and German websites hosted on its Internet servers located in its headquarters in Seattle (Dodge 2001b).

5.4 Geographic Language

Geography matters not only for examinations of Web contents and the applications of the Internet to finances and commerce, but for Internet language and its organization as well. The very flow of electronic information is more abstract than that of capital. Whereas capital flows can be measured quantitatively by a limited number of convertible currencies, it is difficult to measure information beyond its electronic volume, and its contents may be endlessly varied. Capital always departs from a specific location and reaches another one, whereas Internet information is transmitted from one or through several intermediary locations, and it may either not be accessed by anybody, or it may reach several locations simultaneously. Moreover, global capital flows undergo local concretization processes through investments in banks, industrial plants, commerce, etc., which imply fusions with local space. Information remains abstract and uniform all along: at its origin, in its flow and in its destination, although it may also be embodied, for example, in the mass media. Communication is, thus, 'being; persons literally occupy the media they use; their existence cannot be separated from these symbolic systems' (Adams 1995).

5.4.1 Space, place and communications media

Information transmission, notably through the mass media, may play a decisive role in the formation and fusion of space and place, as has been shown

133

for films (Hopkins 1994) and television (Meyrowitz 1985, Adams 1992). 'Virtual geography then is the study of place as ethereal space and its processes inside computers, and the ways in which this space inside computers is changing material place outside computers' (Batty 1997, p. 340). Contemporary information flows are interrelated with experienced space. They reflect an increased curiosity about other places and spaces. Such curiosity may be translated into activity in physical space, such as tourism, which on its part, may bring about additional information flows (see Lash and Urry 1994, p. 309).

The role of computers in mediation processes between the local and the global is different from that of television broadcasting. Computer communications is personal in both the creation and maintenance of social contacts, as well as in the selection of information, compared to monolithic television (Adams 1992). Computers are instantaneous information machines, permitting the storage, processing, retrieval and global transmission of information. These capabilities have yielded global information networks, notably the Internet, which 'have come to be experienced as *places* where we network: a networld' (Harasim 1993, p. 16).

It is important to distinguish between space and place *in* film, television and computerized information, on the one hand, as opposed to these media *as* space/place, on the other. Landscape in films was defined 'as a filmic representation of an actual or imagined environment viewed by a spectator' (Hopkins 1994, p. 49). The process of transforming real landscapes into artistic ones was termed by Benjamin (1986) *mechanical reproduction*. However, space in computers is more complex in its reproduction and expression than in films and television. Space can be reproduced in computerized information in rather simulated and schematic forms, such as maps, so that there are no definitive versions of machine-dependent received and delivered landscapes (see Tivers 1996). As we shall see below, space further serves as a language for the very use of computer information networks, so that language and space become unified rather than separate as was argued for television (Adams 1992).

Television *as* place was defined as:

(1) *a bounded system in which symbolic interaction among persons occurs (a social context), and*
(2) *a nucleus around which ideas, values, and shared experiences are constructed (a center of meaning).*

It is obvious that these two conceptions of place are closely related: social life is founded on shared meanings and meanings are created through social life; each constructs the other.

(Adams 1992, p. 118).

In computerized information transmission, the interaction among persons is both symbolic and interpersonal. These two interrelated aspects of the medium (computers) *as* place are tied with the representation *of* space and place. The simulation of space and the use of spatial language are forms of representation which constitute social formation expressed through the construction of virtual communities, as well as through a possible emergence of a virtual sense of place. The latter, on their part, may reinforce the use of geographical language and symbols, so that a signification of place and space takes place (see Hopkins 1994, pp. 51–52). Thus, virtual space production may involve three dimensions: simulation through pictures or maps, metaphor and language, and virtual communities through interpersonal networked communications.

5.4.2 Simulation

Simulation refers to two or three-dimensional presentations of non-local territories and landscapes, in the form of pictures, maps, or short films. 'Simulation is clearly of major import, indeed simulation is the essence of virtuality' (Batty 1997, p. 343). Simulations, in the form of physical and virtual representations, are easier to grasp and learn than other forms of information (Fabrikant and Buttenfield 2001). However, simulated territories on cyberspace may be perceived as the territory itself, rather than the common separation between the two (Kwan 2001a). Simulations constitute an integral part of media such as motion pictures and television. However, in using the Web, the 'importing' of simulated landscapes is controlled by the viewer, regarding choice and type of information.

Prior to the emergence of the Internet a person had to visit immovable places, whereas computer images allow the viewer to remain geographically static and the 'places' to become moveable. Still, despite the growing sophistication of multimedia transmission, simulated landscapes cannot replace the uncontrolled and unmediated physical presence in other locations, but they do permit the transmission of geographical images without the friction of distance, time and costs involved in physical visits. Simulated geographical

135

images may faithfully represent transmitted landscapes. However, sometimes they may bring about a homogenization of distinct places and landscapes, such as in the standardized street maps of hotel reservation systems (e.g. the *Magic* system).

Some commentators have argued that simulated landscapes may possibly be integrated in the near future into *mirror worlds*, the technology for which exists already. 'You will look into a computer screen and see reality. Some part of your world... will hang there in a sharp cool image, abstract but recognizable, moving subtly in a thousand places' (Gelernter 1991, p. 1). If the proper technology is developed and socially adopted, then such places may not only enhance the functioning of virtual communities, but they may permit production and business from home or from any other sites. Under such circumstances real places and virtual ones may fuse with each other.

5.4.3   Language and metaphor

The English language was the dominant language on the Web in 2000. Some 78% of all websites were in English, climbing to 96% among e-commerce websites, whereas only 50% of all Internet users were native English speakers (Lyman and Varian 2000). The Web has reinforced the status of English as the contemporary most international language. Language is an important ingredient in place making (Tuan 1991), and also vice versa: space can become an important element in language construction. Space serves as a metaphor in computer networks and programs in two ways: verbally, through the use of geographical terms, and graphically through the adoption of place icons and symbolic landscapes. Geographical terms are, for example, *traveling, navigating, cruising, surfing*, or *home* and *site* (in the Internet system) (see Schrag 1994). Geographical notions are also used to describe one's position on a network, through phrases such as: *see you online!*, *let's meet online*, or just *I'm here* (Harasim 1993). The common phrase for major information transmission systems *the information superhighway* is also geographical.

Graphically, the use of place icons and symbolic landscapes is rapidly approaching Gibson's (1985) science-fiction cyberspace, envisioning three-dimensional urban landmarks replacing text and icons. Symbolic places and landscapes have been used as opening screens for various past or present versions of standard computer programs, network communications, and websites (e.g. MS-Office, Apple's eWorld, Magic Cap, ImagiNation Network, Urbanpixel), so that they constitute a *geographic interface* (Schrag 1994).

136

Symbolic places are also used as guiding structures for social networks (e.g. MOOs), which are organized along neighborhoods, buildings, rooms, etc.

Like simulations, metaphors make it easier to learn and grasp. Spatial metaphors are attractive since they are well known to computer users from their daily lives, are simple to use, and make things seem tangible (Schrag 1994, Graham 1998). Further, like simulations, it may turn out difficult to separate metaphors from social realities (see Graham 1998). Spatial language must be kept extremely homogenized and simplified in order for it to be understood by people of different cultures and linguistic languages. This rather shallow global spatial language may be in conflict with much richer domestic spatial languages. It seems, however, that the seemingly declining importance of location in global cyberspace is coupled with increasing importance of geographical language and symbols for instantaneous computer communications. Thus, when space has a somehow more diminished significance, such as territory and distance on cyberspace, it receives an important role as interface, medium, and basic common denominator in virtual images.

5.4.4 Virtual communities

Computerized global networks may develop into placeless social communities, or *virtual communities* (Rheingold 1993a, Mitchell 1995), turning the networks into a new social space (Harasim 1993), or into an *electronic agora* (Mitchell 1995). These communities are based on shared interests of their members attached to *symbolic territoriality* only (Lemos 1996). They may, thus, amount to an *extensibility* of human beings across distance (see Adams 1995). Communities without a spatial anchoring, such as religious ones, are not new (see e.g. Halbwachs 1980, p. 136). However, social bonds in virtual communities may not necessarily be as strong as in dispersed religious communities, since joining and leaving may be much more flexible. Virtual communities do not necessarily replace geographic communities, based on spatial proximity of participants. They rather extend social contacts, similarly to the extension permitted by letter writing or telephone calls (Wellman and Gulia 1999).

Computerized networks may partially replace space as a mediator and context for the emergence and maintenance of human relations, yet space and spatial language play an important role in the functioning of such social networks. Some global networks may develop around an initial location, (e.g. the San Francisco-based WELL network) (see Rheingold 1993b); others, such as MOOs, may also be organized around a symbolic city, implying centrality and agglomeration in the volume and intensity of communications to specific 'rooms', 'buildings', or 'neighborhoods' (see Schrag 1994). Such a network structure may also imply differentiated times and codes of usage between business meetings in 'meeting rooms', and chats in 'cafes' (Harasim 1993).

Spatial structuring of virtual communities constitutes a fusion of local and global spaces. Concepts and ways of behavior, shaped and originated locally, are fused with global social networks which set their norms of behavior accordingly. Global networks and communities may be banned or censured by governments. Thus, China used until recently to prevent its Internet users from free contacts with other countries (Lou 2001), whereas Saudi Arabia, Singapore, Korea, and Iran impose restrictions on global or on comprehensive import of information through the Web (*The Economist* 2001a).

5.5 Conclusion

The relationships between Internet content and geography are rather complex. On the one hand, the Internet, the supposedly most geography-free information system, is based on geographic metaphors and simulations. On the other hand, however, the content of the Web is extremely varied and location-free, in that it does not necessarily reflect the economic specializations of the information-producing places. From a third angle, major content applications, finances and commerce, are heavily location-dependent for their operations, whether it be on telecommunications infrastructure, distribution, or human exchange.

Contents and flows of information may be ubiquitous and universal, but human experience remains spatialized, as reflected in the geography-rich Internet language. Economic activities performed through the Web and through other electronic information systems, do not require space for stores, but they still require locations and space for headquarters, warehouses, distribution centers, and of course, for server farms.

TRANSMISSION

6

'IT and telecommunications networks are thus becoming, in a very real sense, the very sinews of our society'.

(Graham and Marvin 2000, p. 71)

Telecommunications systems have been developed during the last three decades into a complex web of various types of transmission lines and exchanges, wireless waves, satellites, etc. It is not the purpose of the following discussion to highlight the geographies of these systems at large, as these have been exposed elsewhere (Kellerman 1993a, Graham and Marvin, 1996; 2001). It is rather attempted here to present the geography of the Internet transmission system, and to highlight it in light of the previously presented geographies of information technologies, information production, and contents of the Internet. The Internet transmission system interlinks with the telephone system, for the *last mile*, or end-connections from telephone exchanges to PCs, as well as for the growing *voice-over-IP* (VoIP) telephone calls and fax transmissions through the Internet, which are interconnected to telephone lines at the receiving party's telephones.

As its name implies, the Internet interconnects many computer networks. It is typified by its complexity, but at the same time also by its openness to connections via the TCP/IP and other communications protocols. These features of the system avoided its collapse when a major junction of the system in downtown New York (Telehouse) was damaged during the terror attack on September 11, 2001.

The Internet transmission system consists of four major layers: the *physical layer*; *data link*; *network layer*; and *transport* (Gorman and Malecki 2001). In a largely simplified way, the *physical layer* constitutes the web of lines connecting computers worldwide, as well as the dialing systems (modems, communications modules). When a call is made from one computer to another, normally to an *Internet Service Provider* (ISP), the data link checks the unique numerical address of the called computer and its location on the network. Then, the network link establishes the best path to move data between the calling and called computers using routers. An interactive session or reliable transport of data is then established. Similarly, connections are established between ISP computers and website hosting computers.

The hardware components of Internet interactions, namely the telecommunications lines, the *colocations* (the 'switchboards'), the *server farms* (website hosts), and obviously users' PCs, all have their own geographies. The business organization of the interaction/transmission system follows these functional lines. Thus, *Transit backbone ISPs* own and manage backbone transit, so that all other transmission services depend on them (e.g. WorldCom). These other transmission services are: *downstream ISPs*, namely the local and regional ISPs (e.g. Tokyo Internet in Japan); *Web hosting companies* specializing in the owning and managing of server farms for website hosting (e.g. Exodus); and *online service providers* which provide the customer access and visualized interaction (e.g. AOL) (Cukier 1998, see also Gorman and Malecki 2000, Grubesic and O'Kelly 2002). Another classification of international ISPs focuses on the geographical range of services distinguishing among global, regional, national and academic IISPs (*TeleGeography 2001*, 2000). The following discussions will focus on the transmission components, whereas user locations will be left for Chapter 8.

6.1 The Internet Backbones

6.1.1 Definitions

Internet backbones were defined as:

> *A set of paths that local area networks (LANs) connect to for long-distance connection. A backbone employs the highest-speed transmission paths in the network. A backbone can span a large geographic area. The connection points are known as network nodes or telecommunication data switching exchanges (DSEs).*
>
> *(NTIA 2000).*

International backbones were defined as:

> *Private data links which cross international political borders, run the Internet Protocol (IP), are reachable from other parts of the Internet,*

*and carry general Internet traffic: e-mail web pages, and most of the
other popular services which have come to define today's Internet.*
(TeleGeography 2001, 2000, p. 102).

The Internet backbones, as the 'highways' of Internet traffic, are typified in
the same way as transportation highways, by speed, accessibility, and connectivity. Like transportation routes, their crucial role is the efficient moving
of the Internet products, information and knowledge, between producers and
consumers (O'Kelly and Grubesic 2002). Again as in transportation networks,
it is not always clear whether demand (information) brings about supply
(backbones), or the other way around, and in a dynamic system such as
the Internet, demand and supply may also emerge simultaneously. Internet
backbones may constitute in most cases independent fiber optic lines, or
they may constitute leased telephone lines. They are frequently installed,
mostly within cities, along right-of-way infrastructures, such as railways and
sewers (*The Economist* 2001a). Their capacity is measured by bandwidth,
normally in million bits per second (Mbps).

6.1.2 The U.S.

In conjunction with its pioneering in the development of Internet technology,
production, and content, the first Internet backbones were constructed in
the U.S. NSFNET, the major pre-Internet network had its backbone completed
in 1988, through a joint effort by IBM, MCI, the State of Michigan, and the
University of Michigan, followed later by the private *Commercial Internet
Exchange* (CIX) (Wheeler and O'Kelly 1999). The backbone networks have
experienced fast growth in capacity, e.g. 420% between 1997 and 1999
(Moss and Townsend 2000). In 1999, the backbone system was dominated by
WorldCom, Sprint, and *Cable & Wireless*, which together accounted for about
55% of the market (*TeleGeography*, 2000). With 41 transit level backbones
in operation in 2001 the system was rather volatile, as far as mergers
and acquisitions were concerned, raising the share of *WorldCom* to 37% of
wholesale backbone traffic (O'Kelly and Grubesic 2002). Following a general
trend in the telecommunications market, backbone companies have begun to
offer additional services, such as telephone and cable TV services (Kellerman
1997, O'Kelly and Grubesic 2002).

Figures 6.1 and 6.2 present the evolution of the Internet networks and
backbones from ARPANET through NSFNET into the Internet, along three
decades 1971–2000. Given the original purposes of ARPANET, as an alternative defense communications system, and as a connection among security-
related research centers, the network in 1971 basically connected several

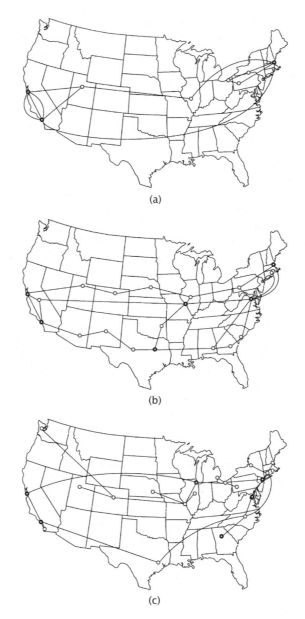

(a)

(b)

(c)

Figure 6.1 ARPANET networks: (a) 1971; (b) 1980; and (c) NSFNET T-3 backbone 1991. Reproduced from (Townsend 2001a) by permission of Pion Limited, London

Figure 6.2 Largest combined intermetropolitan links on 41 Internet backbone networks, 2000. Reproduced from *Economic Geography* (Malecki 2002) by permission of Clark University

sites in and around Boston, Los Angeles and San Francisco (Townsend 2001a). In 1980, Washington, as well as several other installations, were added into the network, the basic geographic contours of which resembled already those of the current backbone system, though major nodes such as New York and Chicago were not yet, or almost not yet, connected. The 1991 NSFNET backbone had these nodes well connected, so that the 2000 network of 41 intermetropolitan systems presents additional connections into Miami, as well as direct connections between Seattle and several cities in Texas.

Figure 6.2 presents the compiled bandwidth on the largest 105 intermetropolitan backbone links in 2000 with at least 5000 Mbps in total bandwidth, but these links comprised only 9.6% of the total of 1100 fiber optic lines connecting U.S. cities (Malecki 2002). The pattern resembles the major economic areas and routes in the U.S., peaking along Megalopolis, as well as between Megalopolis and the Midwest, notably Chicago, and in California (between Los Angeles and San Francisco). These are also the areas

of highest competition among backbone firms, with over 20 of them competing on each line (Malecki 2002). The basic infrastructure for the flow of
information is thus similar to those for the flows of people and commodities, namely expressways and airline routes. The strong connection between
Seattle and Texas, notably Dallas is striking, reflecting connections between
two high-tech centers.

City rankings on the backbone system are normally figured through the
total bandwidth of intermetropolitan backbones serving them. However,
using another pertinent measure, the number of links to other metropolitan
areas, may yield different rankings (Table 6.1). The ten leading metropolitan
areas had a high total share of the system, about 59% of the links, and
about 50% of total bandwidth. Though this concentration is slowly diffusing,
notably westward, the links were still concentrated (O'Kelly and Grubesic
2002). The 1999 data rank Washington D.C. first in both the number of links
and the total bandwidth, but the dominance of the city in the number of links
is higher than in total bandwidth, attesting to its central role as a capital city.
Washington D.C. is also a center of high-tech industry, with more high-tech
workers than Federal government employees (Grubesic and O'Kelly 2002,
Moss and Townsend 2000). The rankings of Dallas and Atlanta are higher in
total bandwidth than in the number of links, which might reflect their central
location rather than their local interconnectivity with other cities. Major cities,
either in population size (New York and Chicago), or in importance in the
information economy (San Francisco) ranked high in the number of links. Los

Table 6.1 U.S. city rankings for Internet intermetropolitan backbones, 1999

Links			Total bandwidth		
MSA/CMSA	**Number**	**% of U.S.**	**MSA/CMSA**	**Mbps**	**% of U.S.**
Washington, DC	233	9.6	Washington, DC	28 370	7.2
San Francisco/San Jose	216	8.9	Dallas	25 343	6.4
Chicago	203	8.4	San Francisco	25 297	6.4
New York	165	6.8	Atlanta	23 861	6.1
Dallas	153	6.3	Chicago	23 340	5.9
Atlanta	131	5.4	New York	22 232	5.6
Los Angeles	121	5.0	Los Angeles	14 868	3.8
Seattle	74	3.1	Kansas City	13 525	3.4
Boston	63	2.6	Houston	11 522	2.9
Houston	62	2.6	St. Louis	9 867	2.6
Total		**58.8**			**50.3**

Sources: Links: Townsend (2001a); bandwidth: Moss and Townsend (2000).

Angeles ranked relatively low, but this may still change, as the system has not yet reached a steady state. We have noted already that Los Angeles became second in domain registration only in 2000. Moss and Townsend (2000) suggest that Los Angeles will receive more extensive connections once the motion picture and television industries will be completely integrated into the Internet. More puzzling is the absence of Boston from the list of high-ranking backbone cities, given its leadership in the high-tech industry.

Table 6.2, thus, shows the changes over the last four years, so that New York reached the top ranking in 2000. It further presents a group of seven metropolitan cities which have persistently topped the list but in different orders: Washington DC; San Francisco/San Jose; Chicago; New York; Dallas; Los Angeles; and Atlanta (Wheeler and O'Kelly 1999, Townsend 2001a, Gorman and Malecki 2001, Malecki 2002, O'Kelly and Grubesic 2002). These cities present a blending of population size, coastal location, centrality, and well-developed information economies. Chicago was found to be the most accessible city on the backbone network (O'Kelly and Grubesic 2002), similarly to its status on the airline network, and with a far better position than its ranking in high-tech and information production. Though the rankings seem to approach the rank order of the population-based urban hierarchy, a regression analysis for the total bandwidth in 32 metropolitan areas in 2000, using population as the predictor, yielded an adjusted *r-square* coefficient of just 0.519, which was higher than an equivalent analysis for 1999 (0.438), but lower than the one achieved for 1998 (0.619) (Malecki 2002).

Table 6.2 Top ten U.S. metropolitan areas in total bandwidth on Internet backbones serving them 1997–2000

1997	1998	1999	2000
Washington	San Francisco	Washington	New York
Chicago	Chicago	Dallas	Chicago
San Francisco	Washington	San Francisco	Washington
New York	Dallas	Atlanta	San Francisco
Dallas	New York	Chicago	Dallas
Atlanta	Los Angeles	New York	Atlanta
Los Angeles	Denver	Los Angeles	Los Angeles
Denver	Atlanta	Kansas City	Seattle
Seattle	Seattle	Houston	Denver
Phoenix	Philadelphia	St. Louis	Kansas City

Source: Malecki (2002). Reproduced from *Economic Geography* by permission of Clark University.

6.1.3 Europe

The structure of Europe, consisting of many countries, with close relations and cooperation among them, blurs the difference between intermetropolitan backbones within and between countries in the continent. Thus, in the year 2000, 75% of the European international backbone capacity was geared for European rather than intercontinental communications (*TeleGeography 2001*, 2000). In 2001, there were 20 backbones in Europe, and 20 additional ones were under construction, compared with 41 in the U.S., though with much larger capacity, as we will see in the next section (Malecki 2002).

London was the only European city connected to all 20 European backbones in 2000 (Malecki 2002). This leading rank, like leaders among American cities, fits its leadership in the information economy, shown already for technology development and information production. London was followed by Amsterdam (19), and Frankfurt, Hamburg, and Paris (18 each), Berlin, Brussels, Dusseldorf, Milan, Munich, and Zurich (17 each). The prominence of German cities fits German leadership in Internet information production. It remains to be seen if Amsterdam's leadership in backbone connectivity will sustain due to its location and extensive international commerce, or whether the expected soon-coming doubling in the number of European backbones will bring about levels of connectivity which will reflect more an urban population hierarchy and/or world-city status of connected cities.

6.1.4 Global view

It is possible to look at the backbone system from a global perspective from three angles: intercontinental, international, and metropolitan. From the intercontinental perspective, Europe is leading by far in terms of total bandwidth, and this trend may become even more striking with the installation of various new European backbones (Table 6.3). This predominance reflects a growing accent on the Internet in Europe, as well as the rather complex national and urban structure of the continent, and the rapidly growing separation from the American backbone system for intracontinental connections. On the other hand, Latin America is the fastest growing continent, though still modest in bandwidth totals. The low international bandwidth in Asia reflects a dependence on the American system, and judging by the European recent trend, this region may possibly see major growth in the development of intra-Asian lines. The African data present the most severe intercontinental digital divide (which will be generally discussed in Chapter 7). Though the total African bandwidth almost doubled between 2000 and 2001, its share declined from

Table 6.3 International Internet bandwidth by continent in Mbps 2000–2001

Continent	2000	%	2001	%	Growth in %
Africa	649.2	0.2	1 230.8	0.1	89.6
Asia	22 965.1	6.2	52 661.9	5.2	129.3
Europe	232 316.7	62.6	675 637.3	66.2	190.8
Latin America	2 785.2	0.8	16 132.5	1.6	479.2
North America	112 222.0	30.2	274 184.9	26.9	144.3
Total	**370 938.2**	**100.0**	**1 019 847.4**	**100.0**	**174.9**

Source: TeleGeography (2001).

Table 6.4 Leading cities in intercontinental backbones by Mbps, 2001

City	Bandwidth
New York	149 989.5
London	85 518.7
Amsterdam	24 479.6
Paris	22 551.8
San Francisco	20 813.6
Tokyo	16 745.5
Washington DC	13 261.2
Miami	11 912.4
Los Angeles	11 227.0
Copenhagen	10 417.0

Source: Adapted from TeleGeography (2001).

0.2 to 0.1% of the world total, due to the tremendous growth in other parts of the world.

The international lines with the largest capacity in terms of bandwidth in 2000 were the ones between London and New York (26 680.5 Mbps), and between London and Paris (24 340.5 Mbps) (*TeleGeography 2001*, 2000). Following, were many additional European lines with much lower bandwidth in each. The heaviest connection between North America and Asia was the line between San Francisco and Tokyo with a capacity of 7550.0 Mbps.

At the metropolitan level New York serves as 'the Internet's most global metropolis' (*TeleGeography*, 2001, p. 1) (Table 6.4). The city was directly connected to 71 countries in 2001, and its backbone capacity is close to double that of the second-ranking city, London. The cities following London ranked much lower. The key role played by the U.S. is demonstrated by the

147

presence of five U.S. cities on the list of the ten leading cities, followed by four European cities, and just one Asian city, Tokyo. Thus, Miami serves as the hub for Latin America, despite its location outside Latin America.

6.1.5   Switching facilities

The discussion so far has focused on telecommunications lines providing Internet connections among cities, countries and continents. The backbone system consists, however, of another major component, switching facilities. These facilities permit interconnection among several backbones, switching traffic from one backbone to another. They further permit access of local traffic into national and international backbones, and the other way around. Switching is technically performed through routers, but requires financial settlements among the involved ISPs. Two major settlement arrangements are frequently used. The first is *peer-to-peer* (P2P), or *peering*, and the second is *transit*, or *customer–provider* relationship (see Malecki 2002, Gorman and Malecki 2000, 2001, Grubesic and O'Kelly 2002). *Peering* is an agreement between two ISPs permitting the transfer of Internet traffic from one backbone to another at a switching facility without charge, assuming that the two-way traffic is similar in volume. Peering agreements typify the relationships among major and veteran backbone or transit ISPs, whereas *transit* agreements imply a fee paid by one ISP to another, e.g. a local ISP paying a transit fee for access to a backbone.

Switching among backbones and local access to and from backbones are performed in *switching facilities*. The first such facilities for data transmission were installed before the emergence of the Internet as an information system, back in 1991. These *Commercial Internet Exchanges* (CIX or IX) were private enterprises providing switching services. The next generation of switching facilities were part of the NSFNET, the *network access points* (NAPs). Four NAPs were built in the mid-1990s in Washington, suburban New York, Chicago, and San Francisco. These were followed by *metropolitan area exchanges* (MAEs). The NAPs were public facilities permitting access and switching, though the routing equipment, as well as the backbones leading to the switching facilities, were owned by ISPs. The number of NAPs has not grown, but the number of MAEs (mostly private IXs) by 2000 had grown to 94 in the U.S., 78 in Europe, 40 in Asia, and several additional ones in other parts of the world (Malecki 2002). Leading cities in Europe in the number of MAEs are London and Amsterdam, complementing their leadership in total bandwidth.

The tremendous growth in all parameters of the Internet industry, namely content, backbones, ISPs, and consumers and applications, has brought about the construction of large numbers of *colocation facilities (colos)*, providing switching facilities at varying scales of operation. These numbered 1415 in 2001, 92% of which were in the U.S., but 25% of the planned ones were outside the U.S. (Malecki and McIntee 2001). The colocation facilities have received different names, such as *telehouses, telecom hotels*, or *Internet hotels*. They have locational requirements similar to those of server farms, and as we will see in the next section many of them actually serve also as such. Thus, proximity to telecommunications hubs is important, as well as sufficient energy sources for their high consumption, so that central city areas, rich in infrastructure are attractive. Colocation facilities may be housed within spacious abandoned buildings, such as garages or supermarkets. The high energy consumption of colocation facilities has been a major contributor to energy shortages in California, the home of many of these facilities (Malecki and McIntee 2001, Malecki 2002). In 2001, the colocation industry was typified by a remarkable over-supply of colocation space, reaching even 50% in leading areas. This was related to the specific and rather selective requirements of ISPs, in terms of safety, telecommunications equipment and backbone connections (*TeleGeography* 2001).

Table 6.5 presents leading U.S. cities in the number of colocation facilities. The six leading cities, New York, Los Angeles, San Francisco, Washington, Dallas, and Atlanta, accounted in 2001 for 36% of the total U.S. colocation

Table 6.5 Number of colocation facilities in leading U.S. cities, 2001

City	Colocation facilities
New York	121
Los Angeles	87
San Francisco-Oakland-San Jose	84
Washington-Baltimore	75
Dallas-Fort Worth	51
Atlanta	43
Boston-Worcester-Lawrence	41
Miami-Ft. Lauderdale	41
Chicago-Gary-Kenosha	36
Seattle-Takoma-Bremerton	36
Denver-Boulder-Greeley	32

Source: Malecki and McIntee (2001).

facilities. This list is similar to that of total bandwidth in city names but not in city rankings (Tables 6.1 and 6.2). The striking leadership of New York, Los Angeles, and San Francisco, leaders in information production, on the one hand, and the absence of Chicago which leads in its backbone capacity but not in information production, on the other, are related to the double service of colocation facilities as switching centers and website hosts (Malecki and McIntee 2001).

The distribution of switching facilities in the U.S. is similar to the distribution of backbones networks, with the exception of Chicago (Figures 6.2 and 6.3). Three inland adjacent states do not enjoy any switching facility: North Dakota, South Dakota, and Wyoming. Townsend (2001b) attributes the concentration of Internet infrastructure in New York to its being the major gateway to Europe, with similar roles played by San Francisco for connection to Asia, as well as by Miami for Latin America.

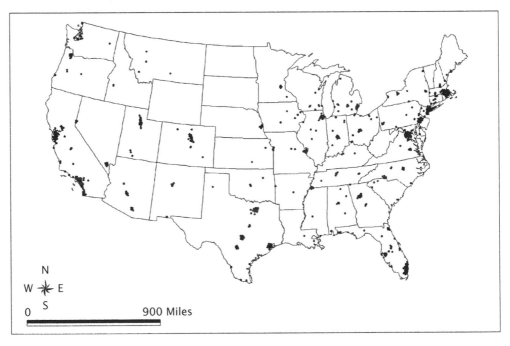

Figure 6.3 Internet switching facilities in the U.S., 2001. (Adapted from Malecki and McIntee 2001)

The distribution of switching facilities in Europe in 2000 was more sparse and uneven, with relatively large numbers of facilities in the United Kingdom, the leading country in European Internet, just a few in other West European countries, and none in East European countries (with the exception of one facility in Russia (Figure 6.4). This pattern may change dramatically with the upcoming installation of more European backbones.

Backbone ISPs have established an extensive system of access facilities, *points-of-presence* (POPs), permitting access of local ISPs and corporations to their backbones. In 2001, nearly 2500 such facilities were in operation in 220 American cities, representing a 200% growth over the period 1996–2000. Obviously, POPs may serve also as colocation facilities at all scales, and vice versa, NAPs and IXs may serve as POPs. Greater San Francisco (including

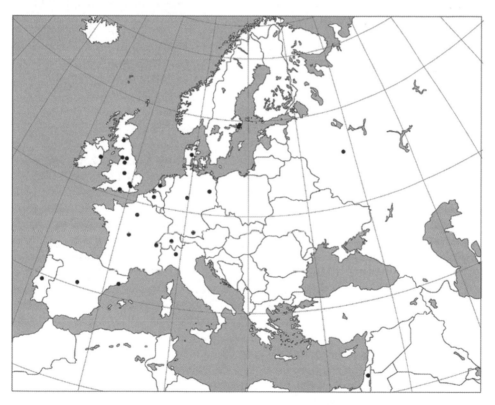

Figure 6.4 Internet switching facilities in Europe, 2000. (Data source: Colosource, 2001)

San Jose, Oakland and the Silicon Valley) led in the number of POPs in 2000 (Grubesic and O'Kelly 2002). Universities worldwide have enjoyed their own Internet communications and switching systems, dating back to the BITNET and EARN academic networks of the 1980s. Even the recently emerging fast academic *Internet 2* network is based on a separate system of national networks, peering through the so-called *gigaPOPS* (Malecki 2002).

Another related telecommunications infrastructure is that of mobile tele-phony towers. These towers have become relevant to the Internet as mobile telephones already transmit data, and are expected to transmit more exten-sive volumes of data (see Chapter 8) (Gorman and McIntee 2002). Some 65 000 towers existed in the U.S. in 2001, 41 204 of which were located in metropolitan areas. Their urban distribution was found to be related to the volume of data traffic, as well as to the location of colocation facilities. The list of leading cities in the location of towers presents, once again, the three leading cities in the Internet economy (Los Angeles, New York, and San Francisco) (Table 6.6). The leadership of Los Angeles over New York may have to do with the larger physical area of the former over the latter, thus requiring more transmission towers. Most of the other leading urban areas lead also in the number of colocation facilities, though in a different order.

6.1.6 Hosting facilities

Website hosting servers are clustered in facilities variously called *server farms, data centers, web hotels*, or *web hosting* facilities. These facilities consist of a large number of servers hosting many websites. An important

Table 6.6 Number of cell towers in leading U.S. MSA/CMSA, 2001

MSA/CMSA	Cell towers
Los Angeles-Riverside-Orange County	2632
New York-Northern New Jersey-Long Island	2216
San Francisco-Oakland-San Jose	1677
Dallas-Fort Worth	1192
Boston-Worcester-Lawrence	1027
Philadelphia-Wilmington-Atlantic City	960
Houston-Galveston-Brazoria	944
Chicago-Gary-Kenosha	897
Washington-Baltimore	858
Denver-Boulder-Greeley	708

Source: Gorman and McIntee (2002).

locational consideration for hosting facilities would be peering or transit availability, so that the website information can be easily transmitted to callers. However, an even more important consideration would be 24/7 maintenance service (Malecki 2002). In addition, such farms require much energy for the servers themselves, as well as for air conditioning. Security is another requirement, so that large abandoned buildings are useful sites, and a combination of a colocation facility and a website hosting facility is possible (see also Chapter 5).

The world capacity of website hosting facilities grows annually by some 50%, and it reached circa 22 million square feet by the end of 2001, despite the technology crisis (*The Economist* 2001a). The business of Internet information management, in the form of website hosting and distribution, is separate from Internet information production and packaging. However, the location of web hotels is not necessarily separate from information production and packaging. Thus, the largest web hosting company, *Exodus*, operated nine facilities in the Silicon Valley, and an additional 35 worldwide. It hosted 49 of the 100 top websites (*The Economist* 2001a). We noted already that the large number of colocation facilities in Los Angeles, despite its relatively low total bandwidth, hints to an extensive hosting of websites in its colocation facilities. Website producers may wish to have convenient access to the servers on which their products reside.

Having a popular website hosted in a single location only may bring about traffic congestion. Thus, several technologies have been developed to permit a more dispersed location of these websites. One set of data replication technologies is *caching*, through e.g. *mirror images* (Gorman and Malecki 2001). Websites are stored in several locations either permanently or as a response to many calls for it. Cambridge, MA headquartered *Akamai*, which is the largest company in this area, operating 11 000 caching servers in 62 countries, i.e. as close as possible to calling users (*The Economist* 2001a). Another website dispersion technology is peering, through which each computer on a network is both a client and a server, so that when a certain website is called for, the system first attempts to locate it in one of the computers connected to it. Such technologies may reduce dependency on costly server farms, but they may bring about heavier traffic (Gorman and Malecki 2001).

6.2 Flows

Generally, data on Internet information flows are unavailable, as ISPs are not required to disclose them. This lack of data is striking when compared to the availability of data on the backbones, or the infrastructure, for the channelling of Internet data flows (see Malecki 2002, *TeleGeography 2001*, 2000). Internet backbones grow faster than flows through them, and it was estimated that in 2002 just 20% of the planned transatlantic backbone capacity will be used for actual traffic (*TeleGeography* 2001). This lack of data on flows is not a new phenomenon, and similar trends were identified for telephony in the 1980s (see Kellerman 1993a). Townsend (2001a) suggested using the geographical patterns of backbone capacity as a surrogate for traffic. However, the rather extensive unused capacity may raise the possibility that current backbone capacities reflect expectations for future demands for data flows rather than present uses.

A Korean study presented Seoul as leading in total information flows, including both incoming and outgoing traffic (Huh and Kim 2001). The share of Seoul in July 2001 was 48.7%, followed by Taejon (18.3%), and Pusan (16.6%), thus presenting a hub-and-spoke pattern. This predominance of Seoul is similar to its leadership in domain name registration reported in the same study (see Chapter 4). The correlation coefficient between the two variables was 0.82, a relationship which assumes that website hosting takes place in cities of their respective domain registrations. The predominance of Seoul was explained by Huh and Kim (2001) as related to the general status of Korea as a developing country, though as far as the Internet is concerned Korea presents high levels of both penetration and use (see Chapter 8). While one may assume a lower dominance of major cities in information flows in developed countries, the share of global cities may still be high, as we have noted already for Tokyo and New York regarding more traditional forms of information flows, such as telephone calls, television and newspapers (see the beginning of Chapter 4).

A still minor application of the Internet network, voice telephone calls (VoIP), is reported as growing fast (*TeleGeography 2002*, 2001). It reached 5.3 billion minutes in 2000, representing a tripling compared to 1999, but still just 5% of telephone traffic. The U.S. was the major source of outgoing calls with a share of some 90%, going primarily to Mexico.

Traffic flows include not only structured and wanted information. The spread of viruses represent 'negative', destructive and unwanted flows. A study of the spread of the *code-red worm* (Crv2) on July 19, 2001 showed that 359 000 computers were infected in less than 14 hours, at the peak of which 2000 hosts were infected every minute (Moore 2001). Of the infected hosts, 43% were in the U.S., 11% in Korea, 5% in China, and 4% in Taiwan.

6.3 U.S. Leadership in Telecommunications

The predominant role of the U.S. in the emergence of the information economy and society has been highlighted so far in various respects, notably information technology and information production. The U.S. has had a leading role even in the transmission of information within countries. In mid-2001, over 80% of international Internet capacity in all continents, except for Europe, still connected directly to an American city (*TeleGeography* 2001). Moreover, until the end of the 1990s, much of the European intranational traffic was transmitted through the U.S. This situation emerged due to the better and cheaper international backbones to the U.S. This pattern has led Cukier (1999a,b) to term the U.S. status as *backbone colonialism*, and the Internet as *U.S.-centric Internet*.

The telecommunications and information economies at large have been characterized by the supremacy of the U.S. in actually all of their operational phases: innovation, production, marketing, adoption and use. However, the growing roles of telecommunications and information worldwide may challenge American leadership in these fields. This is especially so, given the growing trends of globalization, which imply transnational economic investments in telecommunications and information, as well as amplified use of, and dependence on, internationally-based telecommunications and information facilities.

6.3.1 The historical development of U.S. leadership

The U.S. has been described elsewhere as a leading core and activity center for almost every aspect of telecommunications (see Kellerman 1993a). It is where most technological innovations in telecommunications have taken place, since and including the inception of the telegraph in 1837, and the telephone in 1876. The U.S. has further become the first information and service economy (Kellerman 1985), in the emergence and prosperity of which telecommunications technologies have played a crucial role. As we shall see in the next chapter, the American society has further been characterized by fast and extensive diffusion and adoption rates for the various telecommunications technologies. This has been shown time and again for transmission media (e.g. the telephone, the television), for networks (e.g. satellites, fiber optics), for transmission systems (e.g. digital equipment), and transmission nodes (e.g. teleports).

The massive adoption of telecommunications technologies and devices in the U.S. has been coupled with their intensive use. Thus, American society has led in the number of telephone calls per subscriber, as well as in call-lengths. The U.S. has further been shown to be leading in international traffic to and from the U.S., in both public and dedicated networks. Private ownership of telecommunications services has existed until the 1980s almost exclusively in the U.S., so that it was in the U.S. where telecommunications services have always been viewed as a mix of utility and business. With the pioneering introduction of competition in telecommunications services in the mid 1980s, the U.S. has turned into a leader in the global reorganization process of the telecommunications industry (see Kellerman 1993a).

The historical and on-going leadership of the U.S. in telecommunications may be attributed to several factors. The initial advantage of the U.S. in the information economy has been related to technological developments during the cold-war era of the 1950s and 1960s, as well as to the American society being an open society, rather than representing a pre-planned trend (Nye and Owens 1996, Castells 1996, pp. 51–52). However, these historical developments have turned, as of the 1980s, into a policy of increased openness and deregulation of the telecommunications and information industries, thus enhancing the global innovative and technological lead of the U.S.

Economically, the private ownership of the telecommunications system, even if monopolistic until the 1980s, coupled with the capitalist structure of the American economy, have provided incentives for continued R&D efforts as well as for extensive uses of the telecommunications system. Geographically, the large territorial size of the U.S., and the development of the U.S. into a nationally integrated economy, have both been assisted by and challenging to the telecommunications industry. Socially, American society has been characterized as a locationally dynamic one, in that the propensity to move to new locations has been higher than in other countries. This geographical restlessness, coupled with the American openness for innovations and new products, has been aided by telecommunications means, and at the same time may have provided an impetus for the continued development of the telecommunications industry (Kellerman 1993a).

The continued predominance of the U.S. in information production and distribution constitutes a process of mutual reinforcement (Figure 6.5). The prolonged American invention and leadership in telephony was coupled with a rapid diffusion and fast pace of innovation in the computer as well as television industries in the 1960s. The telecommunications revolution of the 1970s marked an integration of computer, telephone, and television technologies (see Kellerman, 1993a). One of the major innovations of the telecommunications revolution, the PC, brought about the development of a

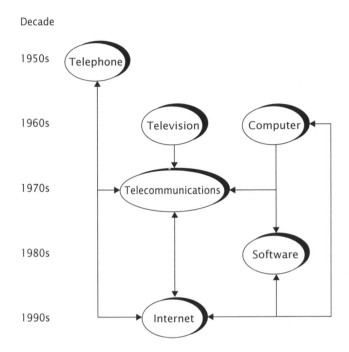

Figure 6.5 American predominance in information media, 1950s–1990s

new industry, namely mass production of software for all PC applications. All these technologies were necessary components for the development and diffusion of the Internet in the 1990s. The global adoption of the Internet and its continued centrality in the U.S. implies continued developments in the previously developed technologies: telephone, television, computer, telecommunications, and software.

The geographical sources of major innovations and inventions in telecommunications were found to be divided between the U.S. (telegraph, telephone, telephone switchboards, television, transistor, semiconductor, cable television, digital switching, picturephone, packet switching, communications satellites, laser and fiber optics, cellular radio), the U.K. (facsimile, television, telex), and Italy (radio) (television was experimented simultaneously in the U.S. and the U.K., which is also true for telegraph, though the latter was patented in the U.S. by Samuel F.B. Morse) (Kellerman 1993a). The senior share of the U.S. (13 inventions) compared to Europe (4 inventions), has been striking throughout the development of telecommunications, and even more so during the post World War II era. It exemplifies the transition from the industrial revolution, which was headed by European nations, to the information revolution dominated by the U.S. (Kellerman 1985).

This traditional leadership of the U.S. was challenged in the 1980s (Cooke and Wells, 1992). Whereas, in 1969, 72% of the patents in communications were American, this share declined to 58% in 1986. It was mainly for Japan to grow in this regard, increasing its share from 7% in 1969 to 20% in 1986. At the same time, production has become relatively cheaper in other parts of the world, whereas the U.S. economy and society moved much faster into the areas of information production and transmission. This American accent on information production and transmission has been coupled with the U.S. leadership in computer development and production, noted already in Chapter 4. Thus, in 1997, the U.S. led in the production of equipment for electronic data processing, and telecommunications (including wireless), whereas Japan led in the production of electronic consumer goods and components (OECD 2000b). The innovation and extensive diffusion of computers in the U.S. have led to American superiority in the area of software, which has become a retail item with the spread of PCs in the 1980s. In the mid-1990s, the U.S. still produced three-quarters of software worldwide, followed by the U.K. and Germany with 10% each, but it was for Ireland to be the largest exporter of software (OECD 2000b).

6.4 Conclusion

The Internet transmission system has been shown to be complex in structure, but flexible in its routine operations, so that the routing of messages between two locations may differ from message to message, using switching points between various backbones, and moving freely among countries, not necessarily using the shortest geographical travel. These complexity and flexibility characteristics proved themselves as major assets of the system during and after the 9/11, 2001 attacks, when a major switching facility in downtown New York was damaged, but the Internet continued operations.

The Internet transmission system may be viewed as an abstract one to some degree, as the fiber optic cables serving the system are telephone lines dedicated for Internet transmission (Townsend 2001b). The *last mile* connection between ISP servers and customers' PCs are in fact telephone lines, so that the Internet and telephony systems are interconnected. By the same token, switching and hosting facilities within the Internet system may be physically interconnected, by sharing the same locations. These complexities may become more extensive with the spread and adoption of broadband transmission (see Chapter 9).

Another complex trend is the expansion of intracontinental backbones. This trend may eventually strengthen the urban hierarchy (Malecki 2002). In the U.S., this trend may bring about more bandwidth into Los Angeles, and in Europe it may cause a decline in the leadership of Amsterdam. On the other hand, the high levels of bandwidth in San Francisco are interwoven with its leadership in Internet technology and information production. The continued growth in the establishment of backbones outside the U.S. and Europe may yield much more complex transmission routes in the future.

Whereas the production of telecommunications equipment spread to cheaper countries, the strength and pioneering of the American market moved to the production and international distribution of information. It was only recently that Europe developed its own backbone systems, and it will take a while before this happens in Asia. We have noted in Chapter 5 the rising role of the U.K. and Germany in information production. It may take longer, however, to see a rather radical decline of the U.S. share in information production, because many Asian and European countries may find it difficult to become global competitors, given the international status of the English language and the obvious advantages of the U.S. in this regard. The U.S. is still the largest producer and distributor of information in the world, both within the country and globally. If information is power, then the U.S. still maintains it at the global scale (see Keohane and Nye 1998).

MEDIA

'The 'World Wide Web' hardly lives up to its name'.

(Warf 2001, p. 6)

This chapter, jointly with the following one, shifts our focus from producers of information to its consumers and at two levels of discussion: information media at consumers' disposal, the topic of this chapter, and consumers' behavior, as well as information consumption per se, the topics of the next one. This chapter will focus first on leading nations in the diffusion and adoption of communications media, and will then turn to the *digital divide*, at various geographical levels: within cities, among regions, and among nations. In the concluding section, the leading areas and those lacking in information media, production and use will be tied into the conceptual framework of the spaces of flows and the spaces of places.

7.1 Leading Nations

One of the major characteristics of communications media has been their wide and reasonably-priced availability to individual customers in developed countries, whether in business or households. This has been true for the non-electronic medium of daily newspapers, as of the nineteenth century, and even more so for the electronic ones in the first half of the twentieth century (telephone, radio, television). The newcomers in the last half of the twentieth century (fax, computers, the Internet, cellular phones, and international direct dialing) have presented similar trends (Kellerman 1999a). The level of adoption and penetration of these media may differ from country to country, and obviously from time to time. The increased number of communications media has brought about media specialization, complementarity and competition.

As we have noted in the previous chapter, the U.S. has been the major source of innovation in communications media, and these media have further widely and speedily spread in it (Kellerman 1993a). The following discussions will comparatively examine the level of penetration of communications media in various countries along time, in order to identify the leaders in each medium, and it will further attempt to identify any possible trends in national leaderships. Moreover, this examination may identify regional or global blocs of countries which lead in the adoption of one or more communications media, and others which may lag behind. This last aspect could be of special interest as recent economic development, rising living standards and globalization, should have eroded differences among developed countries in the three major cores (North America, the European Community, and the Pacific Rim). It may well be, however, that other factors, such as cultural traditions and operational communications systems and regulations, may bring about differences in levels of adoption and penetration of communications media.

7.1.1 Relative advantage in telecommunications at the national level

The relative advantage of cities and regions enjoying well developed telecommunications infrastructure, as well as high levels of adoption rates of communications media, has been noted in several regards. On the one hand, telecommunications systems require urban *spatial fixity* for the very location of their infrastructure, but a well developed telecommunications infrastructure may simultaneously facilitate both urban concentration and decentralization processes of economic activities and population (see Kellerman 1993a, Graham and Marvin 1996, 2001). In addition, telecommunications infrastructure and services may promote economic specialization, and they may further provide channels for globalization efforts (Kitchin 1998, Castells 1996, Morley and Robins 1995). At the regional level, areas relatively rich in telecommunications infrastructure and enjoying high penetration levels of communications media may benefit from numerous dimensions of the technology. Telecommunications is a distributional technology as well as a transactional one for both products and services. It further enables production integration and control, and as we have already noted previously (Chapter 5), it enables the separation of capital location from production location (Gillespie and Williams 1988).

What about the national level? It is somehow implied that developed nations share similar levels of adoption rates of communications means (e.g. Cronin *et al.* 1993, p. 677; Morgan, 1992). The data presented in the following sections will present significant differences in the adoption rates of communications means among developed nations. These differences are of significance, since means of telecommunications are important for the production of goods and services at the national level, in a similar way to their importance at the urban and regional levels. At the consumption level, in both households and businesses, the availability of telecommunications means implies the emergence of a *technological potential*, providing more possibilities for information consumption, through interaction, purchases, sales, entertainment, learning, etc. (Silverstone 1993). Increased information consumption, may imply not only a higher life quality, but it may also bring about mutual reinforcement between the household sector and the production one. Obviously, increased use of information through telecommunications means or a higher demand may lead to larger supply, and the other way around.

The following examination and analysis are based on country data in four different years: 1975, 1986, 1995, 1999. These intervals trace the increasing role of electronic information media in the form of computers, fax, direct international dialing, and eventually the Internet. For each year the seven leading nations for each medium were identified. Data for all years and for all communications media were collected for one variable, namely the penetration rate (e.g. numbers of lines, calls, copies or appliances) per 1000 population (with the exception of international calling). The data sources were authoritative international organizations or data collectors: The United Nations, the World Bank, the International Telecommunication Union, Siemens, and NUA. Each of the next four sections will present the data for one year, followed by attempts at explanations of the longitudinal trends.

7.1.2   Leading nations in 1975

Table 7.1 presents the leading nations in the adoption of communications media available at that time, namely the telephone, newspapers, radio, and television. Global leadership was divided in 1975 between the U.S. and Sweden. The first led in the two electronic mass-media: radio and television; whereas the second led in the rather old newspapers, as well as in the newer telephone. Adoption rates for three out of the four media (television, telephone and newspapers) were similar, over 500 per 1000 population.

163

Table 7.1 Leading nations in the adoption of communications media, 1975 (rates per 1000 people)

Rank	Telephone	Newspapers	Radio	Television
1	Sweden 504	Sweden 572	United States 1882	United States 571
2	Switzerland 371	Japan 526	Canada 959	Canada 411
3	United States 366	Norway 412	New Zealand 876	Sweden 352
4	Canada 350	Switzerland 402	United Kingdom 699	Australia 337
5	Denmark 317	Australia 394	Australia 630	United Kingdom 320
6	New Zealand 308	United Kingdom 388	Hong Kong 574	Denmark 308
7	Finland 264	Denmark 341	Japan 465	Finland 306 Germany 306

Sources: Newspaper, Radio, Television: United Nations (1978). Telephone: Siemens (1975). Reproduced by permission of Bellweather Publishing Ltd.

The rate for radios was almost four times higher compared to the other three media, so that on average there were almost two radios per person in the U.S. This high rate may be related to the high level of car ownership in the U.S., assuming that most cars were equipped with radios. The rates for radio adoption were higher in each rank compared to the rates of other media in the same rank, probably for the same reason. Radio use does not require literacy. Transistor technology turned radios into small, portable and cheap appliances which do not require any infrastructure at the user end. Radio listening further permits users to be engaged simultaneously in other activities, such as driving or working.

In the three electronic media (telephone, radio and television) the rates for the second leading nations were much lower compared to the leading nations in each category, reflecting a strong leadership of both the U.S. and Sweden. Canada followed the U.S. in the two electronic mass-media: radio and television, though as far as radios were concerned the Canadian rate was about one-half of the American one. The two North American countries ranked only third and fourth for telephones, and it was for yet another small European country, Switzerland, to rank second.

Altogether, the lists of high-ranking countries were composed of several geographical groups of countries: the four Scandinavian nations, except for radios in which Finland ranked eighth; North America, except for the rather veteran medium of newspapers; the Pacific Rim, through four countries, namely Australia, New Zealand, Japan, and Hong Kong; and two European nations, namely the U.K. (in radio and television) and Switzerland (in telephones and newspapers). Major European countries were missing, both larger ones, France, Germany (except for television, competing with Finland as sixth), and Italy, as well as smaller ones, such as Belgium and the Netherlands.

7.1.3 Leading nations in 1986

Table 7.2 presents the leading nations in the adoption of the major communications media as in Table 7.1, with the addition of international calling minutes, since the availability of reasonably priced international direct dialing spread during the late 1970s and 1980s. Global leadership was divided in 1986 between four rather than two nations. The U.S. continued its leadership in the adoption of radio and television, and Sweden continued its leadership in the adoption of telephones. However, Sweden lost its leadership in newspapers to Japan. Switzerland, a small country which serves as an *international junction* for commerce, banking, international organizations and transportation, ranked first in international calling.

Adoption rates in 1986 were much more varied than in 1975, but the rates for radio continued to lead. Again, in the three electronic media (telephone, radio and television) the rates for the second leading nations in each category were much lower compared to those of the leading nations, reflecting a strong leadership of the U.S., Sweden, and Switzerland. Canada lost its second rank in the adoption rates of electronic media, demonstrating a wider geographical spread of high ownership rates beyond North America. Switzerland kept its ranking as second for telephones.

Altogether, the lists continued to be composed of the same geographical groups of countries: the four Scandinavian nations, North America, and the Pacific Rim, with the addition of the emerging economic powers of South Korea and Singapore. Among major European countries, France and Germany appeared in rather low ranks for radio and telephone adoption, respectively. The appearance of small European countries which serve as major international junctions for commerce, banking and international organizations is striking (Switzerland, Austria, Belgium, Netherlands), notably in international

165

Table 7.2 Leading nations in the adoption of communications media, 1986 (rates per 1000 people)

Rank	Telephone	Newspapers	Radio	Television	Intl. calling minutes per subscriber
1	Sweden 633	Japan 562	United States 2126	United States 813	Switzerland 255
2	Switzerland 501	Finland 535	Australia 1259	Japan 585	Belgium 146
3	Denmark 497	Sweden 521	United Kingdom 1157	Canada 546	Hong Kong 123
4	United States 495	Norway 501	Finland 992	United Kingdom 534	Norway 120
5	Canada 470	United Kingdom 414	South Korea 952	Australia 472	Austria 119
6	Finland 446	Switzerland 392	France 896 New Zealand 896	Switzerland 411	Singapore 118
7	Germany 419	Austria 365	Canada 877	Sweden 393	Netherlands 99

Sources: Newspapers, Radio, Television: United Nations (1990, 1996); Telephone: Siemens (1986); Intl. Call Minutes; International Telecommunication Union (1994). Reproduced by permission of Bellweather Publishing Ltd.

calling. In this communications medium two similar Pacific Rim small junction countries appeared as well (Hong Kong and Singapore), so that no large country was on the list!

7.1.4 Leading nations in 1995

In 1995, the number and variety of communications media which diffused into both households and businesses increased significantly. Thus, Table 7.3 presents the leading nations in the adoption of the major communications media presented in Table 7.2, with the addition of cellular phones, fax, personal computers and the Internet. Global leadership has expanded as well, so that it was divided in 1995 between six nations rather than four in 1986 and two in 1975. The U.S. continued its traditional leadership in the adoption of radio and television, and it led also in the adoption of fax devices, so that altogether it led in three out of nine communications media. Sweden continued its traditional leadership in the penetration of telephones, which was now coupled with leadership in cellular phones, two devices that provide similar services and partially share the same infrastructure.

Leadership in other devices was divided among four nations, all of which were small countries in the Pacific Rim, Central Europe and Scandinavia. Singapore took over from Switzerland in international calling, and Hong Kong took over from Japan in newspapers (since no data were available for newspaper circulation in Hong Kong for 1975 and 1986, it might well be that Hong Kong had led the list already in earlier years as well). Switzerland led in the adoption of personal computers, whereas Finland led in the penetration of the Internet.

Adoption rates in 1995 became even more varied than in 1986, but the rates for radio ownership continued to lead. In fact, the rates for radio and television penetration in the first-ranking nation, the U.S., slightly declined, compared to 1986, presenting saturated markets. On the other hand, the low rate for fax penetration, in the same leading country, presents an early stage in the diffusion process, notably into households. Another striking change was the more than doubling of the adoption rate for international calling, coupled with a change of the leading nation. The leading rate for telephones, on the other hand, in the same leading nation, Sweden, grew only modestly, presenting once again saturated markets under competition with cellular telephony.

The rapid globalization of the diffusion of communications media was expressed by the small differences in most media between the first and second ranking countries, so that the leadership of the first ranking country became less dominant, except for radio and newspapers. Canada gained back its rank as second in the adoption rates of television, and the U.S. ranked second in telephones, personal computers and the Internet.

The lists continued to be composed of the same geographical groups of countries, despite the addition of several communications media: the four Scandinavian nations, North America, and the Pacific Rim. Among European nations, the appearance of small countries which serve as major international junctions continued to be strong (Switzerland, Austria, Belgium, Netherlands), notably in international calling. This was also true for the two similar Pacific Rim countries (Hong Kong and Singapore), so that in 1995, once again, no large country was on the list for international calling! Among major European countries, France strengthened its appearance.

7.1.5 Leading nations in 1999

The variety of communications media in 1999 was identical to that of 1995, permitting a full temporal comparison (Table 7.4). Global leadership

Table 7.3 Leading nations in the adoption of communications media, 1995 (rates per 1000 people)

Rank	Telephone	Newspapers 1992	Radio 1993	Television	Intl. calling minutes per subscriber	Cellular phones	Fax	Personal computers	Internet
1	Sweden 681	Hong Kong 822	United States 2120	United States 776	Singapore 541	Sweden 229	United States 54	Switzerland 348	Finland 543
2	United States 627	Norway 607	Australia 1290	Canada 647	Hong Kong 516	Norway 224	Japan 48 Denmark 48	United States 328	United States 313
3	Denmark 613 Switzerland 613	Czech Republic 583	United Kingdom 1146	Australia 640	Switzerland 403	Finland 199	Hong Kong 46	Australia 276	Norway 278
4	Canada 590	Japan 577	Denmark 1035	Japan 619	Austria 240	Denmark 157	Sweden 37	Norway 273	Australia 221
5	France 558	Finland 512	South Korea 1013	United Kingdom 612	Belgium 239	Hong Kong 129	France 33	Denmark 271	New Zealand 217
6	Norway 556	Sweden 511	Finland 996	France 579	Netherlands 180	United States 128	Netherlands 32	New Zealand 223	Sweden 211
7	Finland 550	South Korea 412	New Zealand 935	Norway 561	New Zealand 179	Australia 128	Austria 31 United Kingdom 31	Netherlands 201	Denmark 147

Source: The World Bank (1997). Reproduced by permission of Bellweather Publishing Ltd.

Table 7.4 Leading nations in the adoption of communications media, 1999 (rates per 1000 people)

Rank	Telephone	News-papers 1996	Radio	Television	Intl. calling minutes per subscriber	Cellular phones	Fax	Personal computers	Inter-net
1	Norway 709	Hong Kong 792	United States 2146	United States 844	Lesotho 1707	Finland 651	Japan 127	United States 511	Sweden 443
2	Switzerland 699	Norway 588	Finland 1563	Latvia 741	United Arab Emirates 988	Hong Kong 636	Germany 79	Australia 469	Canada 428
3	Denmark 685	Japan 578	United Kingdom 1435	Japan 719	Singapore 719	Norway 613	United States 78	Switzerland 462	Norway 413
4	Sweden 665	Finland 455	Australia 1378	Canada 715	Hong Kong 703	Sweden 583	Hong Kong 58	Sweden 451	United States 394
5	United States 664	Sweden 445	Denmark 1318	Australia 706	Ireland 573	Italy 528	Norway 50	Norway 447	Finland 381
6	Canada 655	South Korea 393	Canada 1047	United Kingdom 652	Namibia 572	Australia 514	Australia 49	Singapore 447	Australia 365
7	Netherlands 607	Kuwait 375	South Korea 1033	Finland 643	Albania 525	South Korea 500	France 48	Denmark 414	Singapore 270

Sources: The World Bank (2001a), Internet (NUA, 2001).

continued to expand, so that it was divided in 1999 between seven nations, compared with six nations in 1995, four in 1986 and just two in 1975. The U.S. continued its leadership in three media: the traditional leadership in the adoption of radio and television to which personal computers were added (rather than fax in 1995). Sweden lost its traditional leadership in the penetration of telephones to Norway. Moreover, Sweden presented a new phenomenon, an absolute decline in telephone lines, probably due to the growing penetration of cellular phones. Simultaneously, though, Sweden further lost its leadership in cellular phones to Finland. Thus, top adoption rates for the two leading types of personal voice communications shifted to two other Scandinavian nations. On the other hand, Sweden, rather than Finland, led in 1999 in Internet adoption rates, with the Internet being another, and newer, form of interpersonal communications medium. The non-appearance of Japan as a leading nation, either in personal computers or in the Internet was attributed to a so-called *keyboard allergy*, showing a gap in this country between a knowledge economy and the possible emergence of a consumer information society (Aoyama 2001).

Fax, on the contrary, presented itself more as a means of business communications, with three major economies leading (Japan, Germany, and the U.S.). The striking leadership of Japan in this medium may reflect the time and language differences between Japan and both Europe and the U.S., permitting business communications and translation (see Kellerman 1993a). Whereas Hong Kong continued its leadership in newspapers, an interesting change could be observed in the leading countries in international telephone calls, which had been, in the past, small junction countries in the Pacific Rim and Europe. Singapore and Hong Kong moved from the first and second positions to the third and fourth ones respectively, and new countries from different parts of the world were added. These are mostly countries which underwent major improvements in their telephone system, and which are dependent on larger neighboring nations for their business conduct. Such were the United Arab Emirates (communicating with Saudi Arabia), and Lesotho and Namibia (communicating with South Africa). Ireland presented its newly acquired status of a telecommunications junction and high-tech hub.

In 1999 the trend of increasing variation in adoption rates continued, but the rates for radio ownership continued to lead. Television rates in the leading U.S. renewed their increase, but the older newspapers declined. Generally, then, the newer media presented rather striking growth levels. The decline in the leading levels in Internet penetration rates might relate to the use of different data sources for 1995 and 1999. For most media the differences in the rates for the first and second ranking nations continued to be small, with the exception of international telephone calling and fax. Although the previously leading world regions continued to be so, the mix of countries expanded, with the addition of East European, African and Middle Eastern nations, attesting to a growing geographical spread in the adoption of communications media.

7.1.6 Factors of leadership

Radio has been by far the leading communications media as far as adoption rates are concerned: throughout the studied period its adoption rates have been much higher than for all other communications media. Telephone, newspapers and television presented varied rates in different years. Given the rapid diffusion of cellular telephony as an alternative telephone technology, one may guess that adoption rates for television will lead over telephones in upcoming years. Television also has a higher potential for penetration compared to newspapers, because its use does not require literacy. Thus radio and the two telephone technologies combined may lead the rankings of adoption rates of communications media. They both permit locational flexibility at varying degrees, and do not require any skills or literacy for their operation.

The Internet is a medium with an extremely fast diffusion process. It took the telephone 74 years to reach 50 million users, and just 4 years for the WWW to reach the same (The World Bank 2001b). The factors for the diffusion and adoption of personal computers, and even more so of the Internet, are much more complex. At the physical level, the diffusion of the Internet was related to telephone infrastructures, which on their part were dependent on previously developed road and railway systems, along which telephone lines were installed (Townsend 2001b). This could well be the case in North America and in Scandinavia, though not equally so in both regions, given the different ownership schemes of public utilities. On another level, the diffusion of the Internet was related to higher levels of education, notably regarding English proficiency (Hargittai 1999). Native English speakers were estimated at 13.9% of the world population in 2001, but their percentage of world online population was 43.0%, or over three times their share of world population (*Global Reach* 2001). The special role of the English language on the Internet could explain high rates of computer and Internet adoption in European countries, especially in small ones, in which English proficiency is accentuated in the educational system.

171

At yet another level, the geographical fragmentation of Internet diffusion was assumed to reflect differences in wealth, technology, and power (Castells 2001, p. 212). At the social level things are more complex, in that opposing social policies may yield high diffusion levels of the Internet, such as the Scandinavian social welfare policies versus Californian individualism (see Castells 2001, p. 278). Explicit national policies regarding IT may also yield high adoption rates, such as in Singapore, Korea, and other Asian countries which consider the Internet as part of their national development agenda (Lou 2001).

It seems that the very existence of most or all of these factors does not necessarily guarantee high levels of Internet adoption, and vice versa, the lack of most of these factors does not necessarily avoid a rapid diffusion of the Internet. Thus Japan, which led prominently in the adoption of fax technology in 1999, presented relatively modest levels of Internet penetration of just 37.2% in December 2000. Another country in the region, Korea, considered to be less developed than Japan, achieved Internet penetration rates of 39.8% one month later, and 46.4% in July 2001 (NUA 2001). Domestic cultures vis a vis work, communications, and leisure may have much to do with Internet penetration levels, so that in Japan, the Internet, and particularly fax, are predominantly business media (see Aoyama 2001), whereas in Korea the Internet has become a home entertainment medium.

Tables 7.1–7.4 have demonstrated the leadership of three major regions in the adoption of communications media, namely North America, Scandinavia and the Pacific Rim. They have further shown the relatively minor role of major European countries. Two countries have shown strong and prolonged leadership, the U.S. and Sweden, both representing two leading regions: North America and Scandinavia. The U.S. led throughout the whole period in radio and television and also led in the relatively new fax, probably because of its low purchasing and operational costs in America. Sweden led until recently in telephone penetration, to which leadership in the similar cellular telephony was added. With the exception of Japan's leadership in newspapers in 1986, as well as in fax in 1999, all the other leaders were small countries in Scandinavia (Finland), Central Europe (Switzerland), and the Pacific Rim (Hong Kong and Singapore), the three latter being international junction countries.

Why was it that these specific regions and countries led in the adoption of communications media? It is important to state here, before going into specific details for each region, that the leadership factors have been different for each region, despite the common thread among them, namely their location within the three global cores (North America, Western Europe and the

Pacific Rim). Though these three regions comprise the foci of global economic wealth, not all the countries in these cores have led in the adoption of communications media. Note especially the modest role played by Germany, France, Italy, and even the United Kingdom. Thus, it is difficult to assume that national wealth will always bring about leading levels of adoption of communications media (see Kellerman 1993a, pp. 134–139). Sweden has enjoyed a global leadership in telephone adoption for many years, until 1999 when it was surpassed by Norway, and this may have led to Swedish leadership in the adoption of cellular telephony. However, Finland surpassed Sweden probably around 1996–1997 in penetration rates of the latter (Telecom Finland 1997). Finland, and later on Sweden, further enjoyed global leadership in the adoption rates of the Internet, while rates for other Scandinavian countries have also been very high. As for newspapers, trends were more complex. In 1975 Sweden led the world and the region, followed by Norway. In 1986 it was for Finland to become second in the world and first in the region, followed by Sweden and Norway. As of 1992, through 1996, Norway became second in the world and first in the region.

The roots of Scandinavian leadership in communications have rarely been touched upon, at least in English language publications, but it was for a Finnish scholar to comment on this leadership that 'it is actually astonishing for countries like Finland and Sweden, not usually so well known for their desire to communicate' (Roos 1994, p. 24). Thus, the Scandinavian leadership in the adoption of communications media may actually run counter to cultural traits of the region. The traditionally high levels of newspaper readership in Nordic countries have been attributed to the large number of 'small and medium-sized local newspapers, with a clear division of labour between them' (Host 1991, p. 281). This in addition to the existence of strong national newspapers (Ostbye 1997, see also Weibull and Anshelm 1992).

The Nordic leadership in telephony and cellular telephony relates to a myriad of rather complex geographical, economic and political factors. The Nordic countries have small population densities, and population is scattered in areas far from major cities. The Swedish major telecommunications company *Ericsson* was founded in the 1870s, and had already begun to manufacture telephones in the 1880s. In Finland, the first telephone was installed in 1877, one year after its invention by Alexander Graham Bell in the U.S. *Ericsson* began exporting before World War I and realized that, in order to keep export prices low, the production price per unit should be kept low as well. This was achieved by low profit margins in the domestic market which resulted in an expanded demand. The Finnish *Nokia* company developed

173

similarly (Roos 1994). State monopolistic service providers cross-subsidized tariffs in the peripheries by higher ones in urban areas. This infrastructure and the high profits of the state monopolies assisted later on in the rapid and reasonably-priced cellular telephony, as well as in high levels of its penetration. It further assisted in the creation of equal technological standards in all Nordic countries, so that the same device could be used everywhere in the region.

The U.S. has been shown to be among the leaders in the adoption of all communications media except for international calls. The U.S. produces about one quarter of international calls and leads the world in this regard, but the number of minute calls per subscriber is relatively low, since most long-distance calling is domestic (Kellerman 1993a). In the adoption of radio and television, the U.S. has topped the list permanently and with substantial gaps between its rates and those of following countries. As already mentioned, the high rates for radio ownership were related to high levels of car ownership. For both radio and television the high rates were also related to the low purchase prices for the two appliances, as well as to the private ownership of broadcasting services in America, which turned the two media, many years ago, into an integral part of the mass-consumption marketing system through commercial advertising. Radio and television thus became, at a very early stage, important components of American lifestyle.

The historical and ongoing leadership of the U.S. in telecommunications may be attributed to several factors. The initial advantage of the U.S. in the information economy has been related to the formation of the United States 'at a moment when a historical void was opened up – a space in between the oral and written traditions' (Carey 1989, p. 3). The invention of cheap paper and mechanical reproduction of words made it possible to transport culture and civilization over large distances, thus permitting the creation of a diversified nation over a large territory. As we have already mentioned in the previous chapter, later technological developments were attributed to the cold-war era of the 1950s and 1960s, as well as to the United States being an open society, rather than representing a pre-planned trend (Nye and Owens 1996). However, these historical developments turned, during the 1980s, into a policy of increased openness and deregulation of the telecommunications and information industries, thus enhancing the global innovative, and technological lead of the U.S.

Geographically, the large territorial size of the U.S., and the development of the country into a nationally integrated economy, have both been assisted by and challenging to the telecommunications industry. Socially,

American society has been characterized as a locationally dynamic one, in that the propensity of residents to move to new locations has been higher than in other countries. This geographical restlessness, coupled with the American openness for innovations and new products, have been aided by telecommunications means, and at the same time may have provided an impetus for the continued development of the telecommunications industry (Kellerman 1997).

The leading role of small countries which serve as international junctions in various respects (e.g. commerce, shipping, banking, international or global business) has increased through the development of regional and global economies. Their leadership was especially noticeable in international calling. Switzerland has traditionally served as an international banking center, followed by two other European nations: Belgium as the headquarters of the European Union, and the Netherlands as a regional and later global maritime trade gateway. In the Pacific Rim, the British Empire turned Hong Kong into a major telephone hub, and developed both Hong Kong and Singapore into major strategic harbors. All these countries, especially Hong Kong and Singapore, are urban entities where adoption rates of telecommunications means, tend to be high. The economic growth of the Pacific Rim coupled with the evolution of the global economy have made Hong Kong and Singapore into complex capital and trade centers both regionally and globally. Singapore has further adopted a national policy towards its turning into a global telecommunications hub (Coe and Yeung 2001).

Of interest is the rather early leading role of Australia and New Zealand, which might be related to their being *new societies*, established by immigrants from Western Europe, similarly to the U.S. and Canada. New societies were typified by their openness to social contacts and innovations, as well as by more intense geographic movements of their members across their territories (see e.g. Elazar 1987). These early and rather social-cultural characteristics were later strengthened by economic growth in the Pacific Rim at large.

As the data for leading countries have shown (Tables 7.1–7.4), generally countries which ranked high in one medium also ranked highly in others. Furthermore, there has existed some geographical proximity among leading countries. These areas either may or may not share with veteran or new *communications traditions*. Thus, small Nordic countries presented similar export-oriented industrial policies, strong state involvement and regional cooperation, without a popular communications culture. North America, another leading region, presented opposite trends, namely reliance on domestic markets, low governmental intervention, and a strong popular

communications culture. International junction countries enjoyed developments in their communications media through their geographical location
and related economic structure, which implied globalization. Thus different geographical spheres of activity, as well as different cultures and state
policies may, within specific mixes of factors, yield high adoption rates of
communications media.

The rather modest, but increasing, leading role of major European nations
was probably related to two processes. First, the monopolistic and governmental structure of most communications services in major European
countries was deeply rooted in these countries, so that it has taken longer
for changes to occur, consequently permitting wider adoption of communications media. Furthermore, regional cooperation through the European
Union has developed relatively slowly, because of the large sizes of the
major countries involved and the more complex interests and traditions they
presented. The recent opening of the European market for full competition
may change the rankings of major European countries dramatically. This
may be coupled, however, with increasing demand for communications and
information services and devices in rapidly industrializing countries in the
Pacific Rim.

7.2 The Digital Divide

7.2.1 Definition

The seemingly potential universality of access to the Internet and its communications and information features have initially brought about some hopes
of its becoming a major equalizing medium across social and economic
strata within cities and countries as well as internationally (Warf 2001). In
fact, however, the Internet, and information access and use in general, have
accentuated the gap between the haves and the have nots. The *digital divide*
has basically been defined as 'the gap between those with and those without
access to information and communication technologies, such as the technologies needed to access the Internet and engage in electronic commerce'
(Paltridge 2001).

Access to information refers mainly to the very existence of proper infrastructures, as well as economic abilities to pay for their use. However, the use
of the Internet is further constrained by the *knowledge gap* (Castells 2001,
p. 258). This knowledge gap may relate first, as Castells noted, to school

systems either being able or not to incorporate the use of the Internet in their curricula. The knowledge gap refers, further, to gaps in society or between societies at large. We have noted in previous chapters the dependency of information production on knowledge (Chapters 3 and 4). In fact, information consumption is too dependent on knowledge. Information production was shown to be dependent on high professional skills in the fields of computing, art, and telecommunications, and these skills are required of a relatively small segment, or in just a handful of cities, in order for information to be produced and transmitted. The geographical concentration of information production amounts, too, to a digital divide. Information consumption, on the other hand, requires much lower skills, but ideally from everybody in order for information to be consumed. These skills are: 'getting connected to the Internet, finding information related to a need or problem, retrieving the information, and using it' (Quay 2001, see also Pelletiere and Rodrigo 2001).

In the widest sense, seven spheres (Cs) have been identified for closing the digital divide (United Nations 1999, p. 63):

- Connectivity – setting up telecommunications and computer networks.
- Community – focusing on group access, not individual ownership
- Capacity – building human skills for the knowledge society.
- Content – putting local views, news, culture and commerce on the Web.
- Creativity – adapting technology to local needs and constraints.
- Collaboration – devising Internet governance for diverse needs around the world.
- Cash – finding innovative ways to fund the knowledge society.

Thus, the digital divide, combining wide differences in both economic and educational infrastructure, may constitute the most striking expression of the gaps between developed and less-developed countries, as well as among regions within countries and cities.

7.2.2 International

The global penetration rate for the Internet in terms of people online reached about 8.5% in August 2001, rising from 7.5% in January 2001, and some 6.1% a year earlier, in August 2000 (NUA 2001). As Figure 7.1 and Table 7.5 show, this rate reflects an extremely wide digital divide among

Figure 7.1 Internet hosts per country, July 2001 (Data source: ISC 2001)

> 100,000
10,000–99,999
1,000–9,999
1–999
0

Table 7.5 Internet penetration and population by continent, August 2001

Continent	Internet users (In millions)	% (of world)	Population (In millions)	% (of world)	Internet/ population ratio
Africa	4.15	0.8	818	13.3	0.06
Asia/Pacific	143.99	28.0	3558	58.0	0.48
Europe	154.63	30.1	727	11.8	2.55
Middle East	4.65	0.9	193	3.1	0.29
North America	180.68	35.2	316	5.1	6.90
Latin America	25.33	4.9	525	8.6	0.57
World	**513.41**	**8.4**	**6137**	–	**0.08**

Sources: Internet: NUA (2001); population: Population Reference Bureau (2001).

the world continents, resembling similar patterns concerning telecommunications at large (see Wilson 2001b, Kellerman 1993a). Simple mapping of the number of hosts per country shows that Africa, the Middle East, and parts of Latin America and Asia, are lagging sharply behind North America, Europe, parts of Asia and the Pacific. More sophisticated mappings, such as the number of hosts per population accentuate the gap even further (see Warf 2001). The ratio between the most advanced region, North America, and the most lagging one, Africa, is enormous: North America was 115 times more connected to the Internet than Africa! Some eleven countries had more than half of their population connected to the Internet in circa mid-2001, all of them in the three global economic cores, with a striking representation by Nordic countries (Table 7.6). At the other extreme are countries in the Internet-deprived regions which at that time did not have a single Internet host or IP address: Benin, Equatorial Guinea, Haiti, Iraq, Liberia, Qatar, Reunion, Sudan, Syria, Zaire, and several island nations (ISC 2001). At the production end of the Internet industry, the gaps are even wider. Thus, the ratio of domain names per thousand population in the U.S. was 25.2 in 2000, compared to just 0.1 in India, that is the U.S. had a ratio 250 times higher than India! (Castells 2001, p. 214).

The international digital divide is not just intercontinental. There are gaps even among OECD countries, such as between the U.S. and Japan (Paltridge 2001), not to mention special cases such as the gaps between Israel on the one hand, and the neighboring Palestinian Authority and Jordan, on the other (Ein-Dor *et al.* 2000). Within developing countries, the connected population are predominantly the urban elites, which may present fast growth trends in upcoming years (Wellman 2001, Castells 2001).

Table 7.6 Countries with over half of the population on the Internet, circa mid-2001

Country	% population on Internet	Date
Sweden	64.4	July 2001
Iceland	60.8	December 2000
United States	59.9	August 2001
United Kingdom	55.3	June 2001
Denmark	54.7	July 2001
Hong Kong	54.5	July 2001
Norway	54.4	July 2001
The Netherlands	54.3	July 2001
Australia	52.5	August 2001
Taiwan	51.9	July 2001
Finland	43.9	August 2000[1]

[1] Probably over 50% in mid-2001.
Source: NUA (2001) based on various sources.

A major reason for the digital divide is access prices, which may turn out to be unaffordably high for low wage earners in developing countries, but have been argued to be significant also to the varying levels of Internet penetration among OECD countries as well (Paltridge 2001). This problem has something to do with the monopoly structure of national telecommunications services (Warf 2001). It may further reflect the activities of foreign countries in rather fragile developing economies (Wilson 2001b). The easy transfer of information over the Internet has accentuated another facet of the knowledge gap. Codified knowledge can be moved from developed to developing countries just like information, but it will lack the proper *contextuality* in the sense of previously acquired education/knowledge and the experience of a reasonable number of people at the receiving end (Johnson and Lundvall 2001).

7.2.3 *Intranational*

The intranational digital divide is of a sociospatial nature, in the wider sense of the term *social*, in that it has emerged along social strata embedded within geographical concentrations, whether within cities or among regions. Intranational digital divides typify nations with well developed information and communications systems, as well as developing countries, in which urban elites enjoy access to information and communications services. The penetration and level of use of the Internet was found to be differentiated along race, ethnicity, minority language, income, education, age, and gender, in both the U.S. and the U.K. (Warf 2001, Castells 2001, Graham and Aurigi 1997). These dividing lines are similar to the rather more general divides between haves and have-nots. Though the digital divide in developed countries is claimed to be narrowing (Wellman 2001), still the quality of Internet infrastructure available in geographical concentrations of deprived groups is of much lower quality than that in affluent areas (Castells 2001).

The social digital divide is spatialized at various scales: through the division between urban and rural areas at the regional scale, and through information-rich and information-poor sections within metropolitan areas. The urban–rural, or center–periphery, gap may narrow down either through regionally endogenous efforts to construct proper infrastructures for the benefit of regional enterprises, or through exogenous efforts by core-located companies searching for rural markets (Richardson and Gillespie 2000). As both the American and European experiences have shown, such efforts have to be aided by governmental assistance (Parker 2000, Grimes 2000). At the metropolitan level, the divide is not just between those with Internet availability and those without. As Graham (2001) has shown, private overlay systems are constructed, and *priority packets* transmitted by or transmitted to prioritized patrons, receive faster transmission, notably at bottle-neck times, so that differences in service, notably speed, are differentiated by class. Similarly, housing complexes for more affluent residents are constructed with built-in superior Internet infrastructure.

Various cities have undertaken efforts to bring about narrower digital gaps. Thus, San Diego has initiated a project calling, among other measures, for outreach to the unwired, support for community call centers, and support of computer ownership (*Mapping a Future for Digital Connections* 2001). Houston has developed a program of free electronic mail (Swartz 2001), and Hong Kong has opened free Internet centers.

7.3 Conclusion

A series of concepts have recently been suggested for the description and analysis of the digital divide. From a spatial perspective, the *splintering* of urban space and urban systems by network infrastructures in both developed and developing countries has been identified by Graham and Marvin (2001). Castells (2001, pp. 240–241) views the divide as leading to a dualism:

emerging from the opposition between the space of flows and the space of places: the space of flows that links places at a distance on the basis

181

*of their market value, their social selection, and their infrastructural
superiority; the space of places that isolates people in their
neighborhoods as a result of their diminished chances to access a
better locality (because of price barriers), as well as the globality
(because of lack of adequate connectivity).*

This dualism implies *marginalization* in the space of places, resulting in various, sometimes compatible, survival strategies: the *informal economy* at the local level; the *criminal economy* at the global level; *separation and succession* at the state level (Castells 2001, p. 269).

Viewing the digital divide within the context of the emerging adoption of communications media in the first part of this chapter, permits viewing the spaces of places and the spaces of flows within an interrelated spectrum, despite the spatial splintering and the simultaneous socioeconomic dualism and marginalization (Figure 7.2). Cities and parts of cities, as well as regions, simultaneously constitute components of both spaces of flows and spaces of places. The more a city or its portion has a higher participation in the new systems of global information flows, the more it is integrated into the space of flows, and vice versa. The space of flows is thus organized in a kind of communications hierarchy, in which higher rankings of cities, as far as global flows are concerned, imply the local existence of lower level media (e.g. local newspapers, local radio and television stations). By the same token, the lower a place is on the space of flows, the higher it is on the space of places, in terms of its constituting an economy and society geared more towards the local and the domestic rather than the global, as well as its specialization in the production of material products rather than globally-relevant information-based services.

Thus, cities that specialize in the production of information, such as Internet websites, television programs, etc. and/or the production of information technology, are highest on the hierarchy of spaces of flows and lowest on that of places. Following are places with high rates of Internet consumption. Ranked lower are places with low quality or even lacking Internet infrastructure, but still having reasonable telephone/fax services. Even lower are places with radio/television reception as main information channels. Lowest are places with little or no electronic communications media, with word of mouth as the major communications medium.

This continuum of media and information production and consumption is splintered in space and it reflects, from top to bottom, declining global information production and thus declining leadership. However, it further implies that almost no place on Earth is totally excluded from the space of

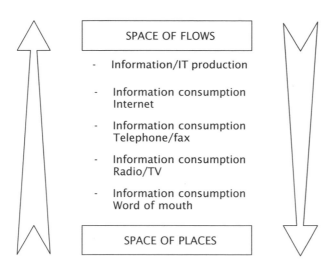

Figure 7.2 The space of flows and space of places: Communications media

flows, even if only weakly or indirectly connected to global sophisticated networks as mere information consumers. Weakly connected places have a more locally focused economy within spaces of places, but they are still connected, and not just through negative aspects such as the criminal economy, but through the sale and purchase of products and services.

Another question is the possible bridging of the digital divide, balancing between domestic and global cultures. Can high levels of telecommunications infrastructure and media adoption and use be reached through the adoption and implementation of proper policies and economic success alone, or does a nation or a city need proper cultural or historical contexts in order widely and usefully to adopt these media? This question is related to the three scenarios and phases proposed in Chapter 2, regarding the relationships between the global and the local: *disembedding*, *phantasmagoria*, and *fusion*. The global and the local may, thus, function as types of separate or even conflicting entities, or they may be partially or fully integrated.

CONSUMPTION

'The Internet is the fabric of our lives'.

(Castells 2001, p. 1)

The *consumption* of information at large, and that of Internet information in particular, normally relates to the availability of information to a user meeting several requirements. First, proper appliances have to be within immediate reach in business and/or at home. The user has to sign up for information service subscriptions which permit constant access to information. This may be true for telephone, fax, and television (notably cable TV) services, as well as for the Internet. Such availabilities of appliances and consumption services do not imply anything on any specific use of information, nor on the quantities consumed, possibly measured through time of connection to an electronic information source, neither does it imply anything on level of use, whether just the passive receiving of information, or being engaged in any processing of information, or in an interactive exchange of information. Data on the consumption and use of information at both the local and individual levels are limited, and obviously tend to focus on consumption rather than on use. If data would permit, it could have been interesting to study the patterns of use of information at its numerous destinations, information which is produced and transmitted from rather few major centers. Are the types of favored information differentiated among users by national geographic location or culture, or perhaps it is differentiated along socioeconomic lines without regard to geographical-cultural location?

This chapter will focus mostly on the consumption and use of Internet information, beginning with some observations on various applications for social purposes, at both the societal and individual levels. It will then move to a highlighting of data on U.S. and Korean cities, comparing consumption data with previously presented data on technology and information production.

The chapter will further expose contemporary efforts to extend and expand the use of the Internet through commercial efforts to identify users' locations, attempting the provision of location-based services and advertisements, as well as efforts to make visual electronic information available, either through wired broadband connections, or through wireless mobile telephones and laptop computers.

8.1 Social Uses of the Internet

One of the rather amazing attributes of the Internet is its extremely wide scope of application. As we have noted already in previous chapters, it carries and moves all sorts of information, and simultaneously constitutes a personal communications system. The Internet serves, furthermore, as a facilitator for the development and functioning of worldwide social networks at an unprecedented level, many of which are loosely structured. The Internet is also a knowledge and information library, and a medium for very structured economic activities and transactions, such as e-commerce and banking.

Reaching all these communications and information systems requires a unique identification of any computer connected to the Internet network, as well as every personal user of the system. A user is identified by an e-mail address, consisting of a personal name or code, located at (@) a hosting computer which is identified numerically through a unique IP-address. These addresses are internationally allocated by IANA (Internet Assigned Number Authority), through three continental authorities: ARIN (American Registry for Internet Numbers) for the Americas; RIPE (Réseaux IP Européens) for Europe; and APNIC (Asia-Pacific Network Information Center) for Asia and the Pacific. These agencies may further delegate responsibility to nationally-based local Internet registries (LIRs) (Dodge and Shiode 2000).

8.1.1 Networking

A network was defined by Castells (2001, p. 1), as 'a set of interconnected nodes'. In the sphere of business management, Castells viewed the Internet as facilitating networks which constitute flexible, adaptable, and coordinating organization tools, permitting a *many-to-many* communications mode on a global scale. He further viewed Internet networking as reflecting globalization, freedom, and telecommunications technology. One may also add to the powers reflected through the success of Internet networking, the contemporary emancipation of women in the labor market, and notably their integration into management positions. Networking being considered a traditional form of female organization (compared to the traditional masculine mode of hierarchical organization). Castells believes that Internet networking is 'outcompeting' and 'outperforming' bureaucracies. However, it does not seem real to assume that hierarchical business organization is about to disappear in the networking era. Though it has been weakened through Internet networking, the fusion of hierarchical and networked cultures and technologies of management have permitted the contemporary management of companies and transactions at an unprecedented scale.

At the social level, Castells pointed to the freedom of expression offered by the Internet in an era dominated by communications media. The versatility of social networks does not permit an identification of a 'unified Internet communal culture' (p. 54). Rather, users are engaged in 'self-directed networking' (p. 55), looking for their own Internet destinations. E-mailing has, thus, been found by the U.S. Census Bureau to be the most popular use in 2000. One-third of adult Americans used the Internet for e-mailing, and 88% among those having home access to the Internet used it for e-mail, followed by 64% using it for information searches (Associated Press, 2001). Among children aged 3–17, 20% used the Internet for e-mailing, and 73% of those having Internet access from home used it for e-mailing, bringing the *Pew Report* to suggest that the Internet has expanded children's friendship networks (Palfini 2001). Some 80 000 *Usenet* groups were reported to exist in 2000, with 8.1 million unique participants exchanging some 151 messages, three times more than in early 1996 (Wellman 2001).

E-mailing and networking have brought various commentators to wonder whether these forms of contact would tend to offset alternative ones, notably the telephone and face-to-face meetings. Furthermore, it was questioned whether the Internet would strengthen or weaken bonds among family members sharing the same household. No conclusive findings have been found in either direction (Wellman 2001, Castells 2001, Sproul 2000). The Internet may increase the number of persons one may be in touch with, located at an unlimited spatial extent, but this does not avoid voice and face-to-face contacts. Similar trends have accompanied the popular and business uses of the telephone at the time (see Kellerman 1993a). Castells (2001, p. 127) went as far as arguing that 'the major transformation of sociability in complex societies took place with the substitution of networks for spatial communities as major forms of sociability', but the very physical existence and basic daily functioning of a person within a specific sociospatial context and *milieu* require local–spatial interactions, so that bridging, adaptation and coordination between the local and the global may be anticipated in well-wired societies, as well as mediation and complementarity between the virtual and the real (see Chapter 2).

8.1.2 Individualism

The Internet may permit an *extensibility* of human beings, in that it relaxes time–space constraints concerning human mobility and activity space, as well as temporal scheduling (Adams 1995, Kwan 2001a,b, see also Kellerman 1989). However, this flexibility may turn out to be of limited nature, given the possible delays in transmission, and the differential time-zones in global real-time interaction. At yet another level, Castells (2001, pp. 130–133) regards the Internet as a material support for the contemporary social trend of *networked individualism* in which networks are created by individuals according to their interests, values, and projects. The Internet, notably when transmitted through mobile telephony, may enhance this social form of activity and organization.

8.2 Internet Consumption in Cities

Whereas the urban geography of Internet production can be assessed through data on the address registration of domain names, which is publicly available, data on Internet penetration at the urban level are much more difficult to obtain, as this information relates to the number of subscribers for the various competitive ISP companies, which in a similar way to telephone companies, normally do not release such information. In the U.S., the Nielsen//NetRatings company surveys the number of households connected to the Internet in major urban markets, and in Korea such surveys are performed by the governmental agency Korea Network Information Center (KNIC). It turns out that data on Internet penetration levels by city are most difficult to obtain.

8.2.1 Penetration rates in American cities

Comparing the U.S. leading *tech-poles* (for information technology production), and leading cities in the number of domains (for Internet information production), to cities leading in Internet home penetration rates for 2000 and 2001 (for consumption), reveals dissimilarities (Table 8.1). Dissimilarities were also found in a previous study using more crude data for 1997–1999 (Kellerman 2000a). Cities that specialize in Internet information production

Table 8.1 The U.S. urban internet economy: technology, production and consumption

Technology			Production			Consumption		
Tech-poles, 1998	Index	MSA	% Domains, 2000		Market	Home penetration, 2000	Market	Home penetration, 2001
San Jose	23.69	New York	12.4		San Francisco	61.0	Portland	69.7
Dallas	7.06	Los Angeles	9.7		San Diego	58.0	Seattle	69.6
Los Angeles	6.91	San Francisco	5.3		Washington	56.1	San Francisco	69.1
Boston	6.31	Washington	3.8		Seattle	55.9	Boston	68.4
Seattle	5.18	Chicago	3.3		Portland	54.0	San Diego	66.2
Washington	5.08	Boston	2.5		Boston	51.7	Washington	65.2
Albuquerque	4.98	Atlanta	2.0		Dallas	47.6	Denver	63.2
Chicago	3.75	Philadelphia	1.9		Denver	47.3	Orlando	61.4
New York	3.67	Seattle	1.9		Atlanta	46.1	Hartford/ New Haven	61.4
Atlanta	3.46				Los Angeles	43.9	Kansas City	61.2

Sources: Technology: DeVol (1999, see Table 3.5); production: Zook (2001a, see Table 4.3); consumption: Internet.com (2001), (Nielsen//NetRatings).

do not necessarily lead in information consumption, or the consumption of Internet information is not dependent on Internet Web information production. In the consumption of information none of the three largest metropolitan areas: New York, Los Angeles, and Chicago, is represented among the top leading areas. Moreover, in 2000 Los Angeles was ranked tenth, falling in 2001 to the 22nd rank. New York was ranked 14th in 2000 falling to the 21st rank in 2001. Mitchelson and Wheeler (1994) in their study of Federal Express data noted that New York 'talked' (or produced information) more than it 'listened' (or consumed information), and 'listening', or information consumption, is determined, among other things, by socioeconomic levels of city residents, which is highly varied in both New York and Los Angeles.

The list of leading cities in the U.S. in Internet penetration levels partially presents cities which specialize in high-tech production, such as San Francisco, Washington, and Seattle. However, leadership in technology production does not necessarily guarantee wide penetration of the Internet into households, as can be attested to by the missing Austin and Los Angeles. The leadership of the San Francisco/San Jose area is most striking, and this area is, thus, the only urban area that leads in all three major spheres of the Internet economy: technology, production and consumption. Moreover, it is for additional and smaller Pacific coast cities to lead in penetration rates (i.e. Portland, Seattle, and San Diego). This leadership may be attributed, on the one hand, to high and more homogenous socioeconomic levels in these cities, as well as to the concentration of high-tech industries in some of them. In addition, it might be related to spillover effects of the San Francisco leadership in uses of the network.

Portland, which led the U.S. penetration rates in 2001 turned out also as leading in that same year in the integration of the Internet into community life (Horrigan 2001). In a comparative study of Portland, Austin, Cleveland, Washington, and Nashville,

> Portland emerges as the leader because its strengths cut across many dimensions. Its combination of technological sophistication, economic vitality, commitment to regional planning, and community engagement, and its existing infrastructure of community nonprofits, make it the city most likely to effectively exploit the Internet for economic and social purposes. Community use of the Internet in Portland extends widely, from neighborhood listservs to community development corporations that are reaching out to low-income people. The business community's active network of angel financiers and entrepreneurs, added to the city's commitment to a new, downtown high-tech center, puts Portland in a good position to compete in the information economy (Horrigan 2001, pp. 2–3).

Whereas New York was portrayed as leading in Web production at large, and Los Angeles was presented as the leader in entertainment applications (Chapters 4–5), San Francisco has been the leader in the use of the Internet as a network. The roots of the city leadership in networking culture may be traced back to the 1960s, with the flower-children revolt against established hierarchies, centered in San Francisco. It so happened later on that most major networks and networking applications of the Internet originated in this city. Such were the sex-oriented Kinky Komputer BBS; the Institute for Global Communication (IGC) advocating the preservation of the environment and world peace; the Homebrew Computer Club, and the Community Memory systems experimenting with computer communications; and the world famous WELL community, a pioneering online community/network (see Castells 2001, pp. 52–53). All these innovations were more social and individual than business oriented. Several of them were also more e-mail than Web oriented, so that the differentiation between information production and consumption was rather blurred. Their spread, adoption and imitation in Western cities implied higher levels of household Internet penetration rates than in business, whereas the opposite might have been the case in major Atlantic coastal cities.

Another trend identified in the Nielsen ratings as reported by *Internet.com* (2001) was the remarkable growth from 2000 to 2001 in penetration rates in older industrial cities such as Philadelphia (40%), Detroit (38%), and Cleveland (35%). Even New York was interesting in this regard (34%), suggesting a rapid move towards more equal penetration rates, as well as towards local Internet markets sized by population. It should be recalled that the August 2001 rate of population online for the U.S. of 59.9% was still low compared to the penetration rates of 95% and higher for household penetration levels for telephone and television. Rising rates of household consumption are related to the pricing of hardware and ISP subscription, ease of operation, and attractive leisure information being offered on the Web.

The 2001 penetration data show much smaller differences among leading cities compared to 2000. Production, as expressed through the location of domains, as well as the innovation and production of technology, are much more concentrated. The future spread of demand for information through high penetration rates in all cities, does not necessarily have to be met by equal diffusions of Internet technology and information production (Pelletiere and Rodrigo 2001).

8.2.2 Penetration rates in Asian cities

The Korean urban pattern of Internet penetration is similar to the American one in the sense that the rate differences among the leading cities are not great, and in that the rankings of Internet penetration differ from population rankings (Table 8.2). However, whereas in the U.S., New York, the largest city, was not among the leading cities in Internet penetration rates, Seoul, the

Table 8.2 Internet penetration rates in Korean metropolitan cities, 2001

City	%	Population rank
Ulsan	63.9	7
Seoul	63.4	1
Daejon	61.8	6
Inchon	61.2	4
Kwangju	56.6	5
Pusan	52.0	2
Daegu	50.0	3

Source: KNIC (2002).

capital and largest city in Korea, is among the leading cities. Seoul was shown in Chapter 4 as dominating the scene of Internet information production in Korea, reaching even global leadership. However, Internet consumption is much more equally distributed, with Seoul being second. Thus information flows in Korea from one center, Seoul, outwards to many dispersed centers, in a similar way to the flows from the three major producing cities in the U.S.

A partially different pattern applies for China, in which Beijing, the leading city in Internet production, led also in Internet penetration, with a percentage of 27.4% in 2001, whereas the penetration rate for the largest city, Shanghai was 17.0% (data deduced for Lou 2001). This pattern may point to socioeconomic differences among Chinese cities, reflecting the level of economic development of China at large.

8.2.3 Use

The use of the Internet is restrained mostly by a user's time-budget, which is obviously a zero-sum property (Kwan 2001b). Lyman and Varian (2000) compared 1992 and 2000 *Bureau of the Census* data on information use in American homes, showing declines in the time spent on the reading of all types of printed materials: newspapers (10%), magazines (6%), and books (4%), coupled with a decline in radio listening (8%). At the same time, watching television registered a small increase (4%), whereas use of the rather new Internet increased tremendously, jointly with video games (126%), and home video (30%). Kwan (2001b) reported that Internet users watch less television and spend less time on shopping, as they are engaged in e-commerce.

Variation in the usage of the Internet may be sizable (see Moss and Townsend 1998). Use has been found to be dependent on education (Pelletiere and Rodrigo 2001), as well as on levels of Internet consumption, so that users in cities with high penetration levels tend to spend more time on the Internet (Table 8.3). Use may have also to do with quality of service. Thus, in Korea, where broadband communications have become a standard Internet service, the average user spends over 18 hours a month on the Internet (see Uimonen 2001). High numbers of hours spent on the Internet were also reported for China (Lou 2001). Use of the Internet in rural areas may be restricted by slower connection times due to lower quality infrastructure (Warf 2001).

Higher penetration levels and extended hours of use may imply higher levels of e-shopping. Thus, in 2001 some 31.7% of the Chinese on the Internet made an e-purchase (Lou 2001). By the same token, in 2000 some 47%

Table 8.3   Consumption and use of the Internet in American cities, 2000

Market	Home penetration rate	Population rank	Monthly average time spent per person (hrs:min:sec) (February)
San Francisco	61.0	5	10:17:44
San Diego	58.0	17	11:15:52
Washington	56.1	4	9:28:58
Seattle	55.9	11	9:11:50
Portland	54.0	22	9:13:21
Boston	51.7	3	8:00:39
Dallas	47.6	9	9:46:52
Denver	47.3	20	10:46:51
Atlanta	46.1	10	7:38:28
Los Angeles	43.9	2	9:09:43

Source: Nielsen//NetRatings (2000).

of Americans having access to the Internet purchased goods and services through the Internet, topped by residents of Washington (60%), Seattle (59%), San Francisco (58%), and Baltimore and Austin with 56% in each (Scarborough 2000). The most popular information items sought were found by this study to be: research/education (51%); news (46%); financial information (36%); local information (30%); sports (26%); and online newspaper (23%). On the other hand, as far as activity on the Internet is concerned, it was reported for the U.S. that doing business with government for activities and transactions such as car or voter registrations, paying fines or enrolling in schools, was performed by 21% of Americans online and was thus more popular than paying credit card bills (15%), and stock trading (10%) (*The Washington Post* 2002).

8.3   Use and Location

The location of Internet consumption and use is not just significant at the national and city levels, but at the level of the individual user as well. Two opposite trends are currently underway in this regard. The first, called geolocation, attempts to identify the geographical location of IP-addresses. Similarly it has become possible to identify the location of mobile phones when a call is made. The second one, wireless information transmission, makes it possible for users to connect to their Internet system or to other information systems, specifically geared to callers on the road, using mobile telephones and wireless laptops, without regard to fixed locations for Internet communications as is required by using a PC.

8.3.1 Geolocation

The purpose of geolocation projects is to identify the geographical location of Internet users, notably of e-commerce customers, as precisely as possible, thus linking the supposedly location-free Internet transactions between buyers and sellers into location-based ones, at least as far as customers' locations are concerned (see Townsend 2000a). Such processes involve three participants: merchants, customers, and regulators, or governments at various scales (local, state, and national).

The customers' approach to having their geographical location automatically identified to potential sellers without their notice, may be mixed. On the one hand, it may make it possible for sellers to offer merchandise that would fit local habits, purchasing power, and needs. It would even permit sending additional information in customers' languages. On the other hand, customers may be exposed to sellers' attempts to sell products with different pricing in different places. Customers may also prefer the anonymity which the Internet permits them, not willing to receive any additional product marketing information. Other groups of customers may not wish to have their identity disclosed for security reasons (Townsend 2000a).

From the merchants' view, geolocation permits the segmentation of marketing in a similar way to practices in physical retailing and direct-marketing. It further permits merchants to meet differing local regulations regarding permitted and prohibited sales of both products (e.g. medications) and services (e.g. gambling).

Regulators are most interested in geolocation as it permits broadening of the sales tax base, which is determined by the location of customers. This may amount to bringing borders back into the Internet (*The Economist* 2001a). Taxing Internet trade would require, however, a high level of geographical accuracy, more than the current circa 90% accuracy at the metropolitan level. Another Internet legal problem, the preservation of intellectual rights, may become easier to monitor through geolocation information (see Chapter 1). Governmental intervention has become a discussed issue, notably following a French ruling in November 2000, prohibiting Yahoo! from selling Nazi memorabilia in France, even though the then available geolocation technologies permitted the country location of customers at an accuracy of just 90%.

A relatively large number of companies and organizations, almost all of which are Silicon Valley-based, have been engaged in recent years in attempts to provide geolocation technologies and solutions (e.g. Quova, Digital Envoy, Infosplit, NetGeo, WideRay, Akamai, NeoTrace, Visual Route, Microsoft, and CAIDA). The developing geolocation technologies are based on several solutions. One major avenue is using the initial data packet sent by a user and containing his/her IP-address and matching it against databases containing IP-address locations. These technologies result in 98% accuracy at the country level, and around 85% accuracy at the city level (*The Economist* 2001a). Other solutions are based on addresses of domain names, as well as on known correlations between the time-length of communications delays and distance between users and website locations (Pierce 2001).

8.3.2 *Wireless information transmission*

Mobile or cellular telephony originally emerged as a complementary communications means to wired telephony, and is currently under intensive R&D development. It is further being rapidly adopted as a means of personal communication. These two trends may turn it into a communications medium by itself. It is not the purpose of the following discussion to highlight wireless telephony, as such, from a geographical perspective (see Townsend 2000b). Rather, it is intended here to discuss wireless telephony as a *portable information machine*, other than voice communications, if for wireless Internet transmission, or for Internet-like information transmission. Kopomaa (2000) viewed mobile telephony at large as placing the sharing of information back into the public sphere, whereas the telephone, radio and television brought it into the home. As such, the cellular phone has become a place in itself, complementing the home and working place and permitting physical mobility, which is the key to its success.

Cellular telephony has probably become the fastest growing communications medium, from about 10 million subscribers in the early 1990s, to 725 million in 2001, or around one cellular telephone for about eight people. Various estimates suggest that the number of cellular lines worldwide will pass the number of telephone lines in the mid decade (2003–2006) (see International Telecommunication Union 2001). In 2001 there were already 35 countries, both developed and developing, which reached this level, and even in Shanghai, China, subscription to cellular telephony surpassed that of wired.

The adoption of mobile information services shows considerable differences among countries, notably in the Pacific. In the relatively early phases of the development of these new services it is still questionable whether

195

these differences are about to change with their continued development and diffusion, or whether they expose differing communications cultures. In Korea, where broadband communications reached the highest household penetration rates, some 4% of cellular subscribers are equipped already with third generation (3G) telephones, permitting users to download full-color pictures. Internet telephones (iMODE), on the other hand, are more popular in Japan, where the charge for this service is low, maybe because Internet telephones serve as substitutes for PC connection, since the installation of PCs and peripheral equipment are space-consuming in small Japanese apartments. Another service, sending and receiving text messages (SMS, for Short Message Service) has gained popularity in Germany, the U.K., Hong Kong, and Singapore, but not in Korea. In the U.S. the popularity of these services seems lower (Uimonen 2001).

Further diffusion of 3G transmission involves the choice between CDMA (Code Division Multiple Access) technology, more popular in Korea, and W-CDMA technology, more popular in Japan. Both are based on 2G CDMA technology, which was in use in about 15% of world mobile networks in 2001, and which is different than GSM (Global System for Mobile Communications) technology, adopted by Europe, and in use by 65% of wireless networks. Thus, it is easier and cheaper for developing countries equipped with CDMA, such as India, Brazil, and Mexico, to adopt 3G technology than it is for Europe and the U.S., which are using other technologies. However, GSM and CDMA have made it possible, technologically, to roam between the two technologies, and 3G wireless telephones are being fitted for use of both technologies which may eventually be adopted in the U.S. (*The Economist* 2002)

The application of mobile telephone technology to the computerized Internet, or for laptop computers used anywhere, has been made possible through the 802.11b communications protocol, called also WiFi (wireless fidelity), or WLAN (wireless local area network). In order for the laptop or the PC to become wireless, it is equipped with an internal antenna, and a proper communications modem, which may be almost as fast or as fast as broadband communications on a PC, and thus much faster than a standard PC modem (see Chapter 9). In addition, a base station is required plus a small external antenna for that station. The system may then operate in one of two ways. Either all subscribers to the base station may connect to each other, or they may all connect to the Internet connection of one of them, or to a connection made available by another wireless party. The technology has so far had two main applications. Wireless services are offered for charge or without charge in places such as airports or coffee shops, or it is organized by citizen

196

groups in various cities in the U.S., sometimes using the available Internet connection of firms and institutions (e.g. schools) after their hours of operation at no cost, since the ISP is paid by bandwidth rather than by time use. Whereas ISPs do not favor this type of connection, citizen groups for wireless networking present the emergence of local networks in a world of global networks, though not for contents sharing, but rather for communications infrastructure.

The use of mobile telephones and their small screens for Internet transmission has not gained popularity in Europe and the U.S., despite huge investments by European carriers bidding for radio spectrum waves for the transmission of Internet data. The idea was to apply full-scale e-commerce into m-commerce (mobile commerce). Mobile communications costs are high, and coupled with security problems and difficulties of use with the small screens of the telephones, this application is still not a success (*The Economist* 2001b). In the U.S., several digital transmission technologies are in use by various operators, compared to the GSM standard adopted all over Europe. The penetration of cellular telephony in the U.S. is still lower than in Europe (Romero 2001) (see Table 7.4)

We noted earlier the geolocation technological effort required to identify the geographical address of PCs transmitting over the Internet. In American mobile telephony, the installation of components to make such an identification possible has become a Federal regulation by the FCC, as of October 2001, attempting to permit easy access to subscribers in emergencies. Though the geographical location of users was already possible earlier, within a certain radius of actual location, the precise geographical identification may be facilitated by the installation of a communications device to satellite-based GPS (Global Positioning Systems). Such connections have paved the way for a whole new range of location-based services (LBS). Users may receive road directions leading them precisely from their location to their destination. They would further be able to receive place-specific commercial information on stores or restaurants within a given range of their location (*The Economist* 2001a,b). The Japanese *J-phone* company introduced its *J-Navi* location-based successful service in October 2000, based on phone numbers and addresses rather than GPS connections. Japanese customers are familiar with the more popular iMODE mobile Internet, so that it remains to be seen whether LBS will gain popularity in the U.S., thus increasing the penetration of mobile telephones. Both in the U.S. and Europe the service may become less successful if subscribers will be flooded with transmitted ads.

197

Urban information is currently arranged by alphabetical order (the telephone *White Pages*), or by business types (the telephone *Yellow Pages*). LBS will call for a geographical ordering of information. This new ordering of urban information may open up new avenues for research in urban geography and planning, both theoretical and applied, attempting an understanding of the full array of business and public facilities from micro-locational perspectives. Location-based services turn mobile telephones into unique information machines as compared to Internet-connected PCs which are general purpose information machines. Like the Internet with its double and interrelated functions as a communications system (e-mail) and information system (the Web), the mobile telephony LBS may yield voice and SMS communications. These services may be integrated with Internet-based e-mail communications. Already in service in Europe are *enhanced messaging services* (EMS) permitting the transmission and presentation of graphic and musical information. This enhanced transmission may be followed in the future by *multimedia messaging* (MMS) for more complex transmissions (*The Economist* 2001b).

8.4 Broadband

Broadband transmission of information over the Internet may apply to any PC, whether used at home or in the office, whereas novel wireless transmission of Internet information has a rather different spectrum of applications mostly outside of homes and offices. Broadband transmission of information implies high-speed transmission of electronic signals from the ISP to the end-user (downstream), and back (upstream). Such fast transmissions are time-saving for Internet communications at large, but are of special importance for the transmission of video information, since broadband permits the transmission of motion pictures at normal speed of picture streaming. It was believed that wide adoption of broadband transmission could generate 'higher growth in productivity rates, as well as new network-based economic activities'. (OECD 2001, p. 4).

Standard telephone modems run at speeds of 28.8 or 56.6 Kbps (kilobytes per second), and by the U.S. Federal Communications Commission (FCC) definition, broadband transmission applies to transmissions at speeds of over 200 Kbps. Thus, ISDN (Integrated Services Digital Network) transmission at the speed of 128 Kbps is not considered broadband, whereas xDSL (Digital Subscriber Line of either A or H technology) or satellite transmissions at speeds of 1.2–1.5 Mbps, as well as cable transmissions at speeds of 2 Mbps are considered broadband transmissions (see Grubesic and Murray 2002).

8.4.1 Diffusion by country

The leading country in terms of broadband penetration into households connected to the Internet is Korea, and neither the traditionally leading country in the Internet industry, the U.S., nor any other Western developed country (Table 8.4). Thus, in February 2001, 57.3% of Korean households who were connected to the Internet, were connected through broadband. This leadership was attributed to the emphasis on education in Korean society, as well as to government support. Culturally, it was related to Confucian society, where home entertainment was preferred. Furthermore, as 40% of Koreans live in apartments it is easier to install xDSL in dense urban structures (Park 2000). It is still too early at the time of writing to assess whether Korea has gained an early bird advantage in the global economy as a result of the wide adoption of broadband. One development related to the widespread adoption of broadband is the leadership of Seoul in domain registration (Table 4.3). A further possible and natural development may be a growing production in the video contents industries, including those of commercials.

Following Korea was the U.S., with just 11.1% (*NetValue* 2001). The U.S. was followed by three other Asian nations (Hong Kong, Singapore, and Taiwan). Interestingly enough, even within the U.S., the leading group in the adoption

Table 8.4 Internet households connecting via broadband, 2001

Country	%
Korea	57.3
U.S.	11.1
Hong Kong	8.1
Singapore	7.1
Taiwan	6.2
France	6.0
Denmark	5.8
Germany	5.0
Spain	3.1
U.K.	3.1
China	0.4

Source: NetValue (2001).

199

of this technology is the Asian-Americans (Castells 2001, p. 256). These patterns may well be related to the cultural patterns identified for Korea. At that same time, in 2001, just 1% of the population in OECD countries enjoyed broadband transmission (OECD 2001). In Korea the penetration of the technology was helped by government, and it is also used by households for social interaction and sending and watching streaming video and audio webcasts produced by users rather than by commercial firms.

The rather slow adoption of broadband, notably in the U.S., has been related to various problems and in several spheres: technology, pricing, and contents. This slow diffusion in the U.S. sheds some light on the difficulties in opening the required gates for a new technology in order to permit its possible universal adoption. The ultimate connection to broadband systems is through fiber optic cables, which require a costly connection from a *central switching office* (CO) to the subscriber's home or office. xDSL technologies attempt to overcome this *last mile* problem by using existing telephone copper wires, originally designed for 4 Kbps voice communications, through equipment installed in COs. The installation of xDSL equipment involves several technical problems, the major one being that this equipment can serve subscribers located within a radius of approximately 5.4 Km only (see Grubesic and Murray 2002). Providers of xDSL services have to negotiate colocation agreements with local telephone companies in order to make use of COs. Lack of such agreements turn the local telephone companies into monopolistic suppliers of broadband services, without competitive prices. Thus, only 3% of the extensive and costly fiber optic backbones connecting COs are in use (Kornbluh 2001).

The relationship between wide broadband adoption and contents is two-way. On the one hand, the slow and costly adoption of broadband prevents the introduction of new applications, such as online doctor visits, high-quality video conferencing, interactive television, as well as business productivity-enhancing applications (Kornbluh 2001). On the other hand, the lack of attractive video broadband content slows down potential customers' adoption of broadband services. The slowness in the introduction of broadband Internet content stems from fear of copyright violations through copying technologies such as *Nepstar* and *MP3*. The concentration of copyrights by several major owners accentuates the difference between an industry of big suppliers, on the one hand, and many and distributed customers, on the other, a situation which requires proper legislation (Lessig 2002).

8.4.2 Penetration rates by cities

In late 2001, some 14–15% of U.S. Internet users had access to broadband; two-thirds of them were located in 25 urban markets, and 44.8% in the top 10 urban markets (Marable 2001, following Nielsen//NetRatings). Such a concentration is typical for early stages in the adoption of a new technology. The leading urban areas present familiar patterns, when compared to consumption patterns (Table 8.5, see also Tables 8.1 and 8.3). New York and Los Angeles led in terms of the number of adopters, thus presenting their largest urban population. Some commentators attributed the large number of subscribers in New York to its dense residential structure, permitting more potential subscribers within the maximum reach of xDSL equipment (Pastore 2001). However, Los Angeles, with a very deconcentrated residential structure, has half the number of New York subscribers, which reflects the population difference between the two cities. San Francisco, which should have had an absolute number of subscribers less than half of those in Los Angeles by its population size, has almost the same number of subscribers. The leadership of this city in the Internet at large is accentuated once again. When the more significant measure of percentage of Internet subscribers connected to broadband service is looked at, West coast cities lead again, without regard to their population rankings. Thus, San Diego tops the list with 35.4%, followed by San Francisco with 30.0%, and Seattle with 26.5%.

Table 8.5 U.S. city rankings for broadband penetration, 2001

Local market area	Population rank	% of U.S. broadband users	% of local Internet users
New York	1	10.2	22.9
Los Angeles	2	5.9	n.a
San Francisco	5	5.8	30.0
Boston	7	4.7	24.2
Seattle	11	3.6	26.5
Dallas	9	3.5	23.8
Chicago	3	3.1	16.0
Philadelphia	6	2.9	20.3
San Diego	17	2.7	35.4
Detroit	8	2.4	19.7
	Total	**44.8**	

Sources: Marable (2001) based on Nielsen//NetRatings.

Table 8.6 Korean metropolitan city rankings for broadband penetration, 2001

City	Population rank	% of local Internet users			
		xDSL	CATV	wireless	total
Kwangju	5	71.6	6.3	0.0	77.9
Ulsan	7	66.5	9.7	0.0	76.2
Inchon	4	67.1	8.7	0.3	76.1
Pusan	2	49.3	24.5	0.0	73.8
Daejon	6	59.2	13.3	0.0	72.5
Seoul	1	50.0	19.5	0.4	69.5
Daegu	3	55.0	13.4	0.3	68.7

Source: KNIC (2002).

In Korea, the leading country in broadband penetration, the urban pattern presents very high rates in the seven leading metropolitan cities, ranging from about 69% to 78% of Internet subscribers in late 2001 (Table 8.6). The ranking pattern is very different from the one for Internet penetration at large (see Table 8.3). Seoul is only the sixth city, whereas peripheral areas tended to adopt broadband with much higher percentages. Also, in the larger cities, CATV seems to be more popular than xDSL as the preferred method of connection and reception.

8.5 Conclusion

The supply side of the introduction of new communications inventions has turned out in recent decades to be extremely productive. However, the very invention and production of new communications media, as well as new modes of consumption and use of information, do not automatically imply a reasonable rate of adoption and use, not even their adoption in principle. A striking example in this regard is the videophone or picture phone, the technologies for which have been developed since 1964, but its daily adoption, other than for e-learning and business videoconferencing, is still very limited, mainly due to social objection. Thus, in addition to its proper pricing, quality of service, and ease of operation, there must be a clear need to be met by a new technology, not restrained by social objection.

It is important to note differences in the popularity of adoption and use of new communications products. This is obviously not just of academic interest, but may shed some light on prospects for future innovations to be adopted within a given social context. This applies first to the national level. Thus, the Nordic countries have been leading the adoption and use of new communications means not just in Europe, but on a global scale. Similarly, broadband transmission for fixed and mobile communications have become very popular in Korea and much less so in Japan, compared to SMS which has become popular in Japan and not in Korea. Such differences may also develop on a regional or local scale, as we noted regarding the leadership of Pacific coast cities in the U.S., concerning Internet penetration rates. Such differences may turn out to be just temporary, through the process of innovation diffusion, so that when the new innovation is eventually universally adopted, the differences may disappear. On the other hand, it might well be that in a world of a wide versatility of communications media, varying levels of the adoption and use of innovations may persist in the long run.

BEYOND

9

'Internet development tends to reinforce rather than destroy urban life and it makes space more rather than less important.'

(Lou 2001, p. 5)

This concluding chapter will first discuss some of the current challenges of the information industry. We shall then follow the opening sentences of Chapter 1, which both suggested and questioned the very existence of a geography of information through two different readings of the title of this volume. This closing chapter will thus begin with a summary of the previous chapters, which hopefully portrayed a geography of information, and more particularly that of the Internet. This exposition will be followed by yet another section, questioning again the very existence of a geography of information.

9.1 Challenges

Wide adoptions of broadband transmission and wireless communications may turn out to constitute keys for a future innovative and dynamic development of the Internet at times of multiple challenges for several components of the Internet/information economy: transmission, innovation, and communications. The transmission industry has been challenged by environmental problems. Internet transmission is typified by an enormous electric consumption of server farms, which have become major electricity consumers, notably in areas of server farm concentration, such as San Francisco with its 16 farms. Such farms require proximity to the Internet information production industry, in case of service needs in the farms by information producers (see Chapter 6). Thus, the *Exodus* San Francisco farm was estimated to consume electricity equivalent to 12 000 houses, leading to high electricity production and the accompanying health problems (Martin 2001). Backup diesel generators for the server farms create pollution. Thus, the city of San Francisco requires special permits from server farms in order to guarantee minimized diesel pollution, as well as maximum electricity efficiency.

The most demanding challenge for the Internet/information industries has been the financial crisis, beginning with declines, as of April 2000, in stock values of companies traded over the *NASDAQ*, which serves as the most important channel for public money mobilization for both technology and computerized information companies. These stock market declines were accompanied by declining demands for computing and telecommunications equipment, as well as the lower availability of venture capital from all sources, notably for the information, so-called .com industry. Castells (2001, p. 89) attributed the volume of the crisis to the increased pace of electronic global stock trading. He further related the crisis to an underestimation of the cost of B2C e-commerce, which involves both virtual and physical facilities; the major investments in information systems towards the anticipated problems with the beginning of year 2000 (Y2K), which involved major investments; political problems in Japan and the U.S.; and unrealistic forecasts by academic analysts (pp. 106–108).

The crisis was preceded in the second half of the 1990s by the emergence of the so-called *bubble economy*, a term related to the extremely high amounts of money involved in investments in high-tech, notably in Internet projects, salaries of professional workers employed in them, as well as sales of innovations and companies at unprecedented prices. This business climate was seen as an attribute of the more general *new economy*, based on computerization and high-tech. However, the major declines in the financial conditions have not reversed the dependence of economic and social life on computers at large, and on the Internet in particular. The crisis may, thus, be viewed as a process of integration of the new economy into the veteran one, in terms of volume of investments in R&D, as well as the operations around company acquisitions and mergers.

The capital of the Internet industry in all of its facets, San Francisco/San Jose, has been hit more than other areas through the financial crisis, and Silicon Valley obviously suffered even more than other parts of the metropolitan area. The second half of the 1990s was typified by soaring real estate prices for both residence and real estate, coupled with increased commercialization, traffic congestion, and socioeconomic gaps (Borsook 1999). By mid-2000 these trends had reversed, reaching 20% vacancies in the .com center of SOMA (South of the Market Area), compared to 0.6% just 18 months earlier (Richtel 2001) However, the region has not lost its major strengths: entrepreneurship, venture capital, and talent, which resulted, in early 2002, in a new round of innovative products, such as new generations of wireless telephony. Venture capital investments which soared to $21 billion in 2000, declined to $6 billion a year later, which is still a high amount, but it was geared mainly to existing rather than new companies (Markoff and Richtel 2002).

In the midst of the global financial-technology crisis, the September 11, 2001 terror attack on the World Trade Center (WTC) in downtown Manhattan occurred. This was one of the geographical cores of the global financial economy, and probably the most equipped information/telecommunications area worldwide. The area experienced telephone service disruptions, either because of infrastructure damage or because of calling traffic jams following the attack. It further suffered from television transmission problems, as major antennas on top of the WTC were destroyed. However, the Internet system continued to function and this despite the damage to the *Telehouse* colocation center, probably one of the largest worldwide. The technology of packet switching, which moves discrete electronic information packets through numerous channels rejoining at the information destination, has proved itself. Thus, the rationale behind the development of the original ARPANET, as an alternative telecommunications system, was put to successful test.

The deeply rooted use of the Internet and its sophisticated infrastructure show that the electronic information industry does not have to reinvent itself, namely to completely change its structure, operations, and innovation processes. The financial and security crises seem rather to lead to a regeneration of the industry, in terms of the expectations of the capital market from continued investments in it, as well as a regeneration of technological products in order for them to fit new demands, something which applies even more to the content industry.

The 9/11 events, the following fear of *Anthrax* and other terror events, have drawn attention to what Graham (2002) described as the interweaving of urban modernity with fragility and vulnerability. Thus, the terrorist network suspiciously performed various stock transactions prior to the 9/11 attack, knowing of its probable financial impacts. Downtown Manhattan has become a symbol of most centralized financial activities globally, networked through information systems. Following the 9/11 events, various financial companies have decentralized part or all of their activities as far away from Manhattan as New Jersey, using these same technologies in

BEYOND

order to facilitate such decentralization. Whereas this move may reflect, in the long run, a further division between front and back offices, the very activity of business negotiation may still require face-to-face meetings in well-established hubs.

Information systems per se, notably the Internet, in both its e-mail and Web components have repeatedly become the target of various forms of attack, mostly through viruses in the e-mail, destruction of websites, or through unauthorized manipulation of databases. Most hit were networks of high technology, financial services, media and entertainment, and power and energy computer networks. It was reported for 2001, that 30% of computer hacking originated in the U.S., followed by France, Korea, and Thailand. However, when adjusted by the number of Internet users, Israel led, followed by Hong Kong and Thailand (SCMP, 2002).

9.2 Geography of Information

In this section we will attempt to outline aspects of a geography of information, notably that of the Internet, following the concluding sections of the previous chapters. The subject matter of this book has been information. Information, in both its broader definition as communicative materials, as well as in its narrower sense of interpreted data or statements, is a major constitutive force in contemporary society, notably through its electronification and fusion through the Internet. Though information is abstract in its very nature, embodied through paper or electronic media as its containers, it has come to have a variety of economic, social, political, and legal aspects for its production, manipulation, transmission, exchange, and use. The geography of information is interwoven into these aspects, despite the abstract and extremely fluid and moveable nature of information, notably in this era of telecommunications.

The information revolution has brought about the emergence of commercial and organizational conglomerates of information production and distribution but, on the other hand, it has also permitted individuals to engage in the production and global distribution of their own information items and products. From a locational perspective, the production and consumption of information has thus, simultaneously, become more flexible (through individual production) and fixed (through concentrations of institutional information production). The immense and constantly growing power of information technology for the handling of information has led somehow to an illusion that information will not be concentrated, so that it may be ubiquitously and freely produced and used. It turns out that information is a social and economic product and resource with its own rules of operation, even under the most fluid conditions. Thus, information constitutes an integral part of the locally unfolding mosaics of power and culture. On the other hand, the ease of global flows of information and the emergence of virtual spatial images turn it into something that transcends local conditions.

Viewing the local and the global, and the real and the virtual, from the standpoint of people as human agents, rather than from the perspective of material or virtual geographical entities experienced or imagined, here and there; then the basis for grasping and functioning of human beings is their knowledge and the information which reaches them and is interpreted by them for use in their daily lives. The most basic contemporary change in the everyday lives of people in developed countries is their constant exposure to information, reaching them from both local and global sources, through the same media and for the same costs. The driving power behind the generation of these geographically varied flows of information might be capital and capitalism, but for urbanites' everyday lives it is information that is the point. The receivers of this global and local, real and virtual information are simultaneously also producers of information which may be transmitted through the same click on the computer keyboard either to a friend or a business across the street, or to a global distribution.

Information systems and information and knowledge per se, therefore serve as crucial factors for the very existence of the virtual and the real, the global and the local, but they further constitute mediators between these worlds. Simultaneity seems to be the name of the game. We experience simultaneous and on-going operations of people, as 'passive' receivers and consumers, on the one hand, constituting at the same time also 'active' producers of information, knowledge and economic activities, within complex dimensions of space and place.

The contemporary handling of information all the way from production through storage, transmission, manipulation and retrieval, implies foremost the application of information technology. Information has become completely integrated with information technology, which, for its part, has become typified by constant innovation based on knowledge. The most important source of knowledge is the uniquely accumulated knowledge, capabilities and experience by each person engaged in R&D, the so-called *tacit knowledge*. For this type of knowledge to be productive and innovative, cross-fertilization amongst holders of such knowledge is crucial, even beyond the boundaries of employing firms. Hence the importance of the geographical location of R&D within specific regions, which permits and fosters knowledge spillover.

The geography of innovation in information technology becomes more complex, given the need for venture capital, as well as additional infrastructure requirements in order for the industry to emerge and flourish. The increased globalization of capital flows, made possible by information technology, permits a wider global distribution of high-tech centers, though the veteran industrial cores are still dominant in the global distribution of high-tech centers. Information technology permits, further, the creation of global networks among these centers, which is essential for centers which house plants owned by MNCs. Therefore, constantly innovative information technologies permit the growing handling of information. At the same time, it is existing information technologies, as well as information per se, which facilitates the development of new generations of these technologies.

The various activities involved in the handling of Web information, from production, through packaging and transmission, may possibly be performed in different cities. However, the geographical convergence among the various phases of information handling, from R&D through production and partially to consumption, is high. The Internet has not constituted a revolutionary new technology, unbased on previous technologies and knowledge. It was rather based on previously existing knowledge, technological as well as artistic, and it permitted the fusion of previously existing media. As such, its development and dissemination has flourished in cities with relevant cumulative advantages.

The geographical concentration of information production in several leading cities, rather than its decentralization, is similar to the continued centralization of global financial activity in various aspects: the previous experience of the leading centers, or the first mover advantage; the global location of customers; the absolute dependence on telecommunications for the transfer of capital and information respectively; and the accumulation of relevant expertise in these centers. Thus, it might well be that the location of the production phase of the Internet industry will disperse, as has really been the case in recent years. On the other hand, leading centers, in both information technology and information production, will continue to pioneer new technologies and approaches, as well as dealing with the heavier and more complex websites, even if there is no significant association between the contents of Internet websites and the more veteran specializations of the cities, such as finances.

The portrayal of the two leading centers of production on the global information space, New York and Los Angeles, may lead to several conclusions regarding the information industry. Information producers which yield electronic products, such as those engaged in capital, FIRE, and the Internet, are concentrated in major centers through all the phases of information manipulation: production, processing and transmission. However, information industries which produce computer-aided material information products, such as books, magazines, disks and cassettes, may locate their manufacturing facilities elsewhere, so that leading cities in the information economy serve as command and control centers, with entrepreneurship, capital mobilization, information production, and decision making concentrated in them. It is for very small areas within leading information cities, such as Manhattan in New York and Hollywood in Los Angeles, to serve as global concentrations of information production, and to enjoy a rather long cumulative advantage.

Several levels of interrelationships may emerge between the information technology and the information production industries. One level of relationship seems most basic, namely client and server relations. The high-tech industry develops the major tools for the Internet information industry, whether it be hardware, software, or telecommunications. Furthermore, the high-tech industry may respond much better and faster to the needs of the information industry, if the two are concentrated within the same cities. The importance of shared tacit knowledge, as well as informal contacts, has been accentuated for both sectors separately, and these may be of importance also for exchanges between the two industries. As Lundvall and Maskell (2000, p. 360) noted: 'innovations reflect a process where feedbacks from the market and knowledge inputs from users interact with knowledge creation and entrepreneurial initiatives on the supply side'. Another added dimension is that the high-tech information technology is a major user of the Internet and its innovations, something which may strengthen the bonds between the two sectors, at the local scene. This type of relationship is most evident in the San Francisco/San Jose area, in which the high-tech industry has become Internet oriented, and well integrated with the leading information industry in the metropolitan area (see Saxenian 1994, pp. 117–118).

There is, however, another level in the relationships between the two sectors. Both have developed along similar organizational and cultural lines.

211

BEYOND

They both require the investment of venture capital, notably in start-up enterprises. Both 'are dominated by the language of individual achievement' (Saxenian 1994, p. 164), channeled many times through small companies, and coupled with dynamism and flexibility, qualities which Saxenian (1994) found to be dominant in the Silicon Valley, and much less so in the Boston area. This distinction may help to explain the superiority of San Francisco over Boston in the contemporary information economy, despite the Bostonian traditional specialization in book publishing. The close cultural and production relations between the technology and information industries may explain the rather large number of domain names in cities which are ranked lower in population size. In other words, in addition to the packaging of information produced elsewhere, local firms in high-tech cities may tend more to go on the Internet, either because of the availability of local expertise, or because of the business atmosphere in high-tech cities.

Despite the seeming benefits from proximity between the technology and information industries, a close relationship between the two industries is neither a geographical nor an economic must for either of them. A high-tech region may specialize in R&D and production of components that do not have direct or immediate implications for the local Internet industry, and websites, on the other hand, may develop with the help of experts in the Internet only. What seems crucial, however, is that *leadership* in the information industry requires the geographical proximity of the high-tech information technology industry, for direct and indirect relations, facilitating constant innovations and pioneering in the computer-aided production and packaging of information.

The relationships between Internet content and geography are rather complex. On the one hand, the Internet, the supposedly most geography-free information system, is based on geographic metaphors and simulations. On the other hand, however, the content of the Web is extremely varied and location-free, in that it does not necessarily reflect the economic specializations of the information-producing places. From a third angle, major content applications, finances and commerce, are heavily location-dependent for their operation, whether it be telecommunications infrastructure, distribution, or human exchange.

The Internet transmission system has been shown to be complex in structure but flexible in its routine operations, so that the routing of two messages between two locations may not be the same, using switching points between various backbones, and having the messages move freely among countries, not necessarily using the shortest geographical path.

212

The U.S. is still the largest producer and distributor of information in the world, both within the country and globally. Adoption and use of information media and new communications products differ from country to country, as well as among cities. The Nordic countries have been leading the adoption and use of new communications means not just in Europe, but on a global scale. Similarly, there has emerged a fast adoption of certain forms of communications media in one country, such as broadband for fixed and mobile communications devices in Korea, but much less so in Japan, whereas SMS became popular in Japan and not in Korea. Such differences may also develop on a regional or local scale, as we noted on the leadership of Pacific coast cities in the U.S., concerning penetration rates for the Internet. Such differences may turn out to determine only temporary differences through the process of innovation diffusion, so that when the innovation is eventually universally adopted, the differences may disappear. On the other hand, it might well be that in a world of a wide versatility of communications media, varying levels of adoption and use of innovations may persist in the long run, reflecting differing cultural preferences.

Cities and portions of cities are integrated into both the spaces of places and the spaces of flows. The more a city or its portion present higher participation in the new systems of global information flows the more they are integrated into the space of flows, and vice versa. The space of flows is thus organized into a kind of communications hierarchy, in which higher ranks imply the local existence in them of lower level media. By the same token, the lower a place is on the space of flows, the higher it is on the space of places, in terms of its being an economy and society geared more towards the local and the domestic rather than the global, as well as its specialization in the production of material products rather than information-based services.

Thus, cities that specialize in the production of information, such as the Internet, television, etc. and/or the production of information technology, are highest on the hierarchy of spaces of flows and lowest on that of places. Following are places with high rates of Internet consumption. Ranked lower are places with low quality or even lacking Internet infrastructure, but still having reasonable telephone/fax services. Even lower are places with radio/television reception as main information channels. Lowest of all are places with little or no technological communications media, with word of mouth as the major communications medium. This continuum of media is splintered in space, and it reflects, from top to bottom, a declining active contribution to global information and thus declining global leadership. However, it further implies that no place on Earth is totally excluded from the space of flows, even if only weakly connected to global sophisticated networks. Weakly connected places have a more locally focused economy or space of places, but they are still connected, and not just through negative aspects such as the criminal economy, but through the sale and purchase of products and services to and from other places.

As we have noted already, San Francisco/San Jose has achieved a special status in the emerging space of flows, given its leadership in almost any aspect of the Internet economy, much higher than its population and global ranking would imply. Townsend (2001a) argues for a city classification of *new network cities*, in which San Francisco/San Jose and Washington, D.C. lead the first tier. San Francisco presents, however, a much stronger leadership than any other city, with the exception of New York and Los Angeles within the U.S., and probably also globally. This has been shown for technology R&D and production, information production, and information consumption (Table 8.1 summarizes these rankings), intercontinental bandwidth (Table 6.4), telecommunications infrastructure (in colocation facilities, Table 6.5; and in cell towers, Table 6.6), and even in adult video (Table 5.5). It has been only in total intermetropolitan bandwidth that Washington leads over San Francisco (Table 6.2), attesting to the role of Washington as the U.S. Federal capital. Information therefore presents concentrations of activity, whether in production or consumption, just as for many other human-economic activities, and these concentrations may locate either in the largest cities or in cities that develop a specialization in this activity.

Several commentators argue for the possible development of virtual states. Such states are supposed to have established ties with a production base located outside their national territory (Rosencrance 1999, p. 209). At yet another level, such a state might be without any geographical territory, so that its sovereignty would be based on the control of information (Edwards 2001, p. 311). Such scenarios and resulting spaces of flows seem to be still far ahead. A recent ranking of countries as information societies, based on infrastructures in computing, the Internet, telecommunications, as well as on some social indicators, has ranked three Scandinavian nations as world leaders: Sweden, Norway, and Finland, followed by the U.S., Denmark, and the U.K. (*The WorldPaper* 2001). The Nordic nations, do not yet seem to be moving towards non-territorial organization, perhaps only for linguistic reasons.

9.3 Geography of Information?

The title of Cairncross's (1997) book *The Death of Distance* has carried with it a notion of the death or elimination of geography altogether in the era of instant global transfer of information through telecommunications. Is geography really dead or meaningless in an information economy and society? And if geography is not dead is there a different geography for information compared with other economic and social activities? Geography constitutes basically two opposing things; namely, convergence and separation. Convergence is usually conceived as location in places, whereas separation in space occurs through distances (see Figure 2.1). Traditionally these were two sides of the same coin called geography. The contemporary ability to move information of all kinds through a single electronic outlet, the Internet, has brought about some separation between these two aspects, as well as some change in their significance as far as information is concerned.

Distance for an end-user may become meaningless in the daily operation of the Internet, if two conditions are met. First, that the user is equipped with the proper infrastructure (e.g. a powerful PC, and broadband ISP connection). Second, that the Web information is within fast access through caching or peering. Such optimal conditions are becoming more and more frequent, and even less than optimal conditions minimize the significance of distance for end-users. Obviously, from the supply side things look different, since overcoming distance is the name of the game for many businesses in the areas of backbone construction, e.g. ISPs, server farm companies and the like.

Location, or places of Internet production, could also become meaningless, since it has become possible to produce websites everywhere. However, location has received a growing significance in the electronic world of information, and for old-fashioned geographical reasons. The production of information has enjoyed the cumulative advantage of major cities in more traditional forms of information, such as book publishing or motion picture production. Furthermore, information is mostly related to other activities, such as finances, entertainment, etc., or to knowledge production, such as universities, all activities well situated in specific places before the introduction of the electronic submission of information. Thus, in the spaces of flows, places are of significant and even growing importance, with one major difference compared with the pre-Internet era; namely, that these places are less viable as independent and separate entities. Cities are networked in a more complex way and more instantly, than ever in the past.

Thus, from an urban geographic perspective, the basic rules of location are still there, namely urban size and urban hierarchy as possible information production predictors, coupled with initial or early mover advantage, as well as with cumulative advantage in information production; and with specialization in information production at large, or in certain types in particular. On the other hand, the personal mobility of information, or perhaps, better, the mobility of personal information, has changed drastically. One such change is e-mail, and the possibility of instantly sending and receiving personal mail from anywhere. The second change is of no less significance, especially in the world of knowledge, namely that a person can carry with them anywhere their personal library which is now fully portable. However, since most carriers of tacit knowledge come and go from a homebase to any close or far location and back, it seems that the basic geographical differences between the various types of information have remained unchanged, even if all are being moved much faster than in the past. Thus, data and information are transferred much more freely than knowledge, notably tacit knowledge, and innovations remain the most concentrated and less moved. Given the interrelationships between the production of knowledge and the production of information, the production of both tends to coincide spatially.

REFERENCES

Abler R F 2000 *Mastering inner space: Telecommunications technology and geography in the 20th century*. Presentation at the Congress of the International Geographical Union (IGU), Seoul, Korea.

Abler R, Adams J S and Gould P 1977 *Spatial Organization: The Geographer's View of the World*. Englewood Cliffs, NJ: Prentice Hall.

Abu-Lughod J L 1999 *New York, Chicago, Los Angeles: America's Global Cities*. Minneapolis: University of Minnesota Press.

Adams P C 1992 Television as gathering place. *Annals of the Association of American Geographers* **82**: 117–35.

Adams P C 1995 A reconsideration of personal boundaries in space–time. *Annals of the Association of American Geographers* **85**: 267–85.

Adams P 1998 Network topologies and virtual place. *Annals of the Association of American Geographers* **66**: 88–106.

Agnew J 1993 Representing space: space, scale and culture in social science. In J Duncan and D Ley (eds) *Place/Culture/Representation* London: Routledge, 251–71.

Agnew J A and Duncan J S 1989 Introduction. In J A Agnew and J S Duncan (eds) *The Power of Place: Bringing Together Geographical and Sociological Imaginations*. Boston: Unwin Hyman, 1–8.

Ahnstrom L 1997 Personal communication.

Alexa Research 2001 *E-Commerce Report Q4 2000*. http://www.alexaresearch.com

Amin A and Thrift N 1994 Living in the global. In A Amin and N Thrift (eds) *Globalization, Institutions, and Regional Development in Europe*. Oxford: Oxford University Press, 1–22.

Angel D P and Engstrom J 1995 Manufacturing systems and technological change: The U.S. personal computer industry. *Economic Geography* **71**: 79–102.

Antonelli C 2000a Restructuring and innovation in long-term regional change. In G L Clark, M P Feldman and M S Gertler (eds) *The Oxford Handbook of Economic Geography*. New York: Oxford University Press, 395–410.

Antonelli C 2000b Collective knowledge communication and innovation: The evidence of technological districts. *Regional Studies* **34**: 535–47.

Aoyama Y 2001 The information society, Japanese style: corner stores as hubs for e-commerce access. In T R Leinbach and S D Brunn (eds) *Worlds of E-commerce: Economic, Geographical and Social Dimensions*. Chichester: Wiley, 109–28.

217

REFERENCES

Appadurai A 1990 Disjuncture and difference in the global cultural economy. *Theory, Culture and Society* **7**: 295–310.

Associated Press 2001 America online? Well, about half. http://www.wired.com/news/culture/0,1284,46582,00.html

Atkinson R 1998 Technological change and cities. *Cityscape: A Journal of Policy Development and Research* **3**: 129–71.

Atkinson R D and Gottlieb P D 2001 *The Metropolitan New Economy Index: Benchmarking Economic Transformation in the Nation's Metropolitan Areas.* Cleveland: Progressive Policy Institute.

Audretsch D 2000 Knowledge, globalization, and regions: An economist's perspective. In J H Dunning (ed.) *Regions, Globalization, and the Knowledge-Based Economy.* New York: Oxford University Press, 63–81.

Audretsch D and Feldman M P 1996 R&D spillovers and the geography of innovation and production. *American Economic Review* **86**: 630–40.

Barlow J P 1994 The economy of ideas. *Wired* **2.03**: 1–17. http://www.wired.com/wired/archive/2.03.

Barthes R 1972 *Mythologies.* A. Lavers (trans.). London: Jonathan Cape.

Barua A, Pinnell J, Shutter J and Whinston A J 1999 *Measuring the Internet economy: An exploratory study.* Center for Research in Electronic Commerce, The University of Texas at Austin. http://crec.bus.utexas.edu

Batty M 1997 Virtual geography *Futures* **29**: 337–52.

Batty M and Barr B 1994 The electronic frontier: Exploring and mapping cyberspace *Futures* **26**: 699–712.

Beaverstock J V, Smith R G and Taylor P J 2000 World-city networks: A new metageography? *Annals of the Association of American Geographers* **90**: 123–34.

Bell D 1976 *The Coming of Post-industrial Society: A venture in Social Forecasting*, 2nd ed. New York: Basic Books.

Ben–David D and Loewy M B 1998 Free trade, growth and convergence. *Journal of Economic growth* **3**: 143–70.

Benedikt M 1991 Introduction. In M Benedikt (ed.) *Cyberspace: First Steps.* Cambridge, MA: MIT Press, 1–26.

Beniger J R 1986 *The Control Revolution: Technological and Economic Origins of the Information Society.* Cambridge, MA: Harvard University Press.

Benjamin W 1986 The work of art in the age of mechanical reproduction. In J G Hanhardt (ed.) *Video Culture: A Critical Investigation.* Layton: Gibbs M. Smith. 27–55.

Biegel S 2001 *Beyond Our Control? Confronting the Limits of Our Legal System in the Age of Cyberspace.* Cambridge, MA: MIT Press.

Bird J, Curtis B, Putnam T, Robertson G and Tickner L (eds) 1993 *Mapping the Futures: Local Cultures, Global Change.* London: Routledge.

Boisot M H 1998 *Knowledge Assets: Securing Competitive Advantage in the Information Economy.* Oxford: Oxford University Press.

Borgmann A 1999 *Holding on to Reality: The Nature of Information at the turn of the Millennium.* Chicago: University of Chicago Press.

Borsook P 1999 How the Internet ruined San Francisco. http://www.salon.com/news/feature/1999/10/28/internet/print.html

Braczyk H J, Fuchs G and Wolf H G (eds) 1999 *Multimedia and Regional Economic Restructuring*. London: Routledge.

Braman S 1989 Defining information: An approach for policymakers *Telecommunications Policy* **13**: 233–42.

Brealey R and Ireland J 1993 What makes a successful financial center? *Banking World*, June: 23–5.

Brooker-Gross S R 1985 The Changing Concept of Place in the News. In J Burgess and J R Gold (eds) *Geography, the Media and Popular Culture*. London: Croom-Helm, 63–85.

Brunn S D and Leinbach T R (eds) 1991 *Collapsing Space and Time: Geographic Aspects of Communication and Information*. New York: Harper Collins Academic.

Brunn S D and Leinbach T R 2000 Nokia as a regional information technology fountain-head. In J O Wheeler, Y Aoyama, and B Warf (eds) *Cities in the Telecommunications Age: The Fracturing of Geographies*. New York: Routledge, 130–42.

Bryson J R, Daniels P W, Henry N and Pollard J (eds) 2000 *Knowledge, Space, Economy*. London: Routledge.

Button K and Taylor S 2001 Towards an economics of the Internet and e-commerce. In T R Leinbach and S D Brunn (eds) *Worlds of E-commerce: Economic, Geographical and Social Dimensions*. Chichester: Wiley, 27–44.

Cairncross F 1997 *The Death of Distance: How the Communications Revolution Will Change Our Lives*. Boston: Harvard Business School Press.

Camagni R 1991 Introduction: From the local 'milieu' to innovation through coop-eration networks. In R Camagni (ed.) *Innovation Networks: Spatial Perspectives*. London: Belhaven, 1–11.

Campbell D 1996 Political processes, transversal politics, and the anarchical world. In M J Shapiro and H R Alker (eds) *Challenging Boundaries*. Minneapolis: University of Minnesota Press, 7–31.

Carey J W 1989 *Communication as Culture: Essays on Media and Society*. Boston: Unwin Hyman.

Castells M 1985 High technology, economic restructuring and the urban-regional process in the United States. In M Castells (ed.) *High Technology, Space and Society*. Beverly Hills: Sage, 11–40.

Castells M 1989 *The Informational City: Information, Technology, Economic Restruc-turing and the Urban-Regional Process*. Oxford: Blackwell.

Castells M 1994 European cities, the informational society, and the global economy. *New Left Review* **204**: 18–32.

Castells M 1996 *The Rise of the Network Society*. Oxford: Blackwell.

Castells M 1998 *End of Millennium*. Oxford: Blackwell

Castells M 2000 *The Rise of the Network Society*, 2nd ed. Oxford: Blackwell.

Castells M 2001 *The Internet Galaxy: Reflections on the Internet, Business, and Society*. New York: Oxford University Press.

Castells M and Hall P 1994 *Technopoles of the World: The Making of Twenty-First-Century Industrial Complexes*. London: Routledge.

CEMA (Consumer Electronics Manufacturers Association) 1999 http://www.cc.org/images/ce_and_us_econ_98/9.gif.

REFERENCES

Chan-Olmsted S M 1998 Mergers, acquisitions, and convergence: The strategic allian-
ces of broadcasting, cable television and telephone services. *The Journal of Media
Economics* **11**: 33–46.
Chandler A D Jr and Cortada J W (eds) 2000 *A Nation Transformed by Information:
How Information has Shaped the United States from Colonial Times to the Present*
New York: Oxford University Press.
Christopherson S and Storper M 1986 The city as studio; the world as back lot: The
impact of vertical disintegration on the location of the motion picture industry.
Environment and Planning D: Society and Space **4**: 305–20.
Ciolek T M 2002 *Networked information flows in Asia: The research uses of the
Altavista search engine and 'weblinksurvey' software*. Paper prepared for the Inter-
net political economy forum 2001, Singapore. http://www.ciolek.com/PAPERS/
weblinksurvey2001.html.
Clark G L, Feldman M P, Gertler M S (eds) 2000 *The Oxford Handbook of Economic
Geography* New York: Oxford.
Coe N M and Yeung H W 2001 Grounding global flows: Constructing an e-commerce
hub in Singapore. In T R Leinbach and S D Brunn (eds) *Worlds of E-commerce:
Economic, Geographical and Social Dimensions* Chichester: Wiley, 145–66.
Cooke P, Uranga M G and Etxebarria G 1998 Regional systems of innovation: An
evolutionary perspective *Environment and Planning A* **30**: 1563–84.
Cooke P and Wells P 1992 Globalization and its management in computing and com-
munications. In P Cooke, F Moulaert, E Swyngedouw, O Weinstein and P Wells (eds)
*Towards Global Localization: The Computing and Telecommunications Industries
in Britain and France* London: UCL Press, 61–78.
Corey K E 2000 Intelligent corridors: Outcomes of electronic space policies *Journal of
Urban Technology* **7**: 1–22.
Cortright J and Mayer H 2001 *High-Tech Specialization: A Comparison of High Tech-
nology Centers* Washington, DC: The Brookings Institute.
Cosgrove D 1984 *Social Formation and Symbolic Landscape* New Jersey: Barnes &
Noble.
Cowan R, David P and Foray D 2000 The explicit economics of knowledge codification
and tacitness *Industrial and Corporate Change* **9**: 211–53.
Cox K R 1992 The politics of globalization: A sceptic's view *Political Geography* **11**:
427–9.
Cox K R 1993 The local and the global in the new urban politics: A critical view.
Environment and Planning D: Society and Space **11**: 433–48.
Crang M, Crang P and May J (eds) 1999 *Virtual Geographies: Bodies, Space and
Relations* London: Routledge.
Crevoisier O and Maillat D 1991 Milieu, industrial organization and territorial produc-
tion system: Towards a new theory of spatial development. In R Camagni (ed.)
Innovation Networks: Spatial Perspectives London: Belhaven, 13–34.
Cronin F J, Colleran E K, Herbert P L and Lewitzky S 1993 Telecommunications and
growth: The contribution of telecommunications infrastructure investment to
aggregate and sectoral productivity *Telecommunications Policy* **17**: 677–90.
Cukier K N 1998 The global Internet: A primer *TeleGeography 1999* Washington, DC:
Telegeography, 112–45.
Cukier K N 1999a Global telecom rout. *Red Herring* **63**: 60–4.

Cukier K N 1999b Bandwidth colonialism? The implications of Internet infrastructure on international e-commerce. http://www.isoc.org/inet99/proceedings/1e/1e_2.htm.

David P A 1990 The dynamo and the computer: An historical perspective on the modern productivity paradox *American Economic Review* **80**: 355–61.

DENIC eG. 2001 *Domain Statistics* Http://www.denic.de/doc/DENIC/presse/stats 2000.en.html.

DeVol R C 1999 *America's High-Tech Economy: Growth, Development, and Risks for Metropolitan Areas* Santa Monica, CA: Milken Institute.

Dodd N 1994 *The Sociology of Money: Economics, Reason and Contemporary Society* London: Polity Press.

Dodge M 2001a Guest editorial. *Environment and Planning B: Planning and Design* **28**: 1–2.

Dodge M 2001b Finding the source of Amazon.com: Examining the store with the 'earth's biggest selection'. In T R Leinbach and S D Brunn (eds) *Worlds of E-commerce: Economic, Geographical and Social Dimensions* Chichester: Wiley, 167–80.

Dodge M and Kitchin R 2001 *Mapping Cyberspace* London: Routledge.

Dodge M and Shiode N 2000 Where on earth is the Internet? An empirical investigation of the geography of Internet real estate. In J O Wheeler, Y Aoyama, and B Warf (eds) *Cities in the Telecommunications Age: The Fracturing of Geographies* New York: Routledge, 42–53.

Doron A 2002 Oracle: 36 million websites on the Internet. *Ma'ariv* January 17 (Hebrew).

Driver S and Gillespie A 1993 Information and communication technologies and the geography of magazine print publishing *Regional Studies* **27**: 53–64.

Duncan J and Duncan N 1988 (Re)reading the landscape *Environment and Planning D: Society and Space* **6**: 117–26.

Dunning J H 2000a *Regions, Globalization, and the Knowledge-Based Economy* New York: Oxford.

Dunning J H 2000b Regions, globalization, and the knowledge economy: The issues stated. In J H Dunning (ed.) *Regions, Globalization, and the Knowledge-Based Economy* New York: Oxford, 7–41.

Eade J (ed.) 1997 *Living the Global City: Globalization as a Local Process*. London: Routledge.

The Economist 1996 *The economics of the Internet: Too cheap to meter?* 19 October: 21–4.

The Economist 2001a *Special report: Geography and the net* 9 August.

The Economist 2001b *What do consumers want from the mobile Internet?* http://www.economist.com/printedition/displayStory.cfm?Story_ID =811994

The Economist 2002 *Mobile telecoms*. January 10. http://www.economist.com/business/displayStory.cfm?Story_ID =930233

Edwards T M 2001 Corporate nations: The emergence of new sovereignties. In T R Leinbach and S D Brunn (eds) *Worlds of E-commerce: Economic, Geographical and Social Dimensions* Chichester: Wiley, 293–314.

Ein-Dor P, Goodman S E and Wolcott P 2000 From *Via Maris* to electronic highway: The Internet in Canaan *Communications of the ACM* **43**: 19–23.

REFERENCES

Elazar D 1987 *Building Cities in America: Urbanization and Suburbanization in a Frontier Society* Lanham, MD: Hamilton Press.

Elkin-Koren N 1996 Public/private and copyright reform in cyberspace *Journal of Computer Mediated Communication* **2**. http://www.ascuse.org/jcmc/vol2/issue2/elkin.html.

Entrikin J N 1991 *The Betweenness of Place: Towards a Geography of Modernity* Baltimore: The Johns Hopkins University Press.

European Commission 1996 *The Information Society* Luxembourg: Office for Official Publications of the European Communities.

Evans P B and Wurster T S 1997 Strategy and the new economics of information. *Harvard Business Review* **75**(5): 71–82.

Eyescream 1997 http://www.eyescream.com/yahootop200.html.

Fabrikant S I and Buttenfield B P 2001 Formalizing semantic spaces for information access. *Annals of the Association of American Geographers* **91**: 263–80.

Feather J 1994 *The Information Society: A Study of Continuity and Change* London: Library Association Publishing.

Federal Communications Commission (FCC) 1997 *Telecommunications Act of 1996* http://ftp.fcc.gov/telecom.

Feldman M P 1994 *The Geography of Innovation* Dordrecht: Kluwer.

Feldman M P 2000 Location and innovation: The new economic geography of innovation, spillovers, and agglomeration. In G L Clark, M P Feldman and M S Gertler (eds) *The Oxford Handbook of Economic Geography* New York: Oxford University Press, 373–94.

Feldman M P and Audretsch D 1999 Innovation in cities: Science-based diversity, specialization and localized competition *European Economic Review* **43**: 409–29.

Feldman M P and Florida R 1994 The geographic sources of innovation: Technological infrastructure and product innovation in the United States *Annals of the Association of American Geographers* **84**: 210–29.

Felsenstein D 1993 *Processes of Growth and Spatial Concentration in Israel's High Technology Industries* PhD dissertation, Jerusalem: The Hebrew University (Hebrew).

Finney M 1998 The Los Angeles economy *Cities* **15**: 149–53.

Florida R 1995 Toward the learning region *Futures* **27**: 527–36.

Friedmann J and Wolff G 1982 World city formation: An agenda for research and action *International Journal for Urban and Regional Research* **6**: 309–44.

Garnsey E 1998 The genesis of the high technology milieu: A study of complexity *International Journal of Urban and Regional Research* **22**: 361–77.

Garreau J 1991 *Edge City* New York: Doubleday.

Geller P E 1996 Conflicts of law in cyberspace: Learning from old media experiences. In P Bernt Hugenholtz (ed.) *The Future of Copyright in a Digital Environment* The Hague: Kluwer Law International, 27–48.

Gelernter D 1991 *Mirror Worlds: Or the Day Software Puts the Universe in a Shoebox. How it Will Happen and What Will it Mean* New York: Oxford University Press.

Gibbs D and Tanner K 1997 Information and communication technologies and local economic development policies: The British case *Regional Studies* **31**: 765–74.

Gibson W 1985 *Neuromancer* London: Gollancz.

Giddens A 1990 *The Consequences of Modernity* Cambridge: Polity Press.

REFERENCES

Giddens A 1991 *Modernity and Self-Identity: Self and Society in the Late Modern Age* Cambridge: Polity Press.

Gillespie A and Williams H 1988 Telecommunications and the reconstruction of Regional Comparative Advantage *Environment and Planning A* **20**: 1311–21.

Gillespie A and Robins K 1989 Geographical inequalities: The spatial bias of the new communications technologies *Journal of Communications* **39**: 7–18.

Gillespie A, Richardson R and Cornford J 2001 Regional development and the new economy. *European Investment and the New Economy* **6**: 109–31.

Global Reach 2001 *Global Internet statistics (by language)* http://www.glreach.com/globstats/index.php3

Goddard J 1990 Editor's preface. In M E Hepworth, *Geography of the Information Economy*. New York: Guilford, xiv–xvii.

Goddard J 1992 New technology and the geography of the UK information economy. In Robins K (ed.) *Understanding Information Business, Technology and Geography* London: Belhaven, 178–201.

Goddard J B 1995 *ICTs space and place: Theoretical and policy challenges*. Paper presented at the Workshop on Informatics and Telecom Tectonics: Information Technology, Policy, Telecommunications, and the Meaning of Space.

Goodchild M E 2001 Towards a location theory of distributed computing and f-commerce. In T R Leinbach and S D Brunn (eds) *Worlds of E-commerce: Economic, Geographical and Social Dimensions* Chichester: Wiley, 68–86.

Gordon D 1988 The global economy: New edifice or crumbling foundations? *New Left Review* **168**: 24–65.

Gorman S P 2001 *Where are the Web factories? The urban bias of e-business location* Paper presented at the Information and the Urban Future meeting, Taub Urban Research Center, New York University http://www.informationcity.org

Gorman S P and Malecki E J 2000 The networks of the Internet: An analysis of provider networks in the U.S.A. *Telecommunications Policy* **24**: 113–34.

Gorman S P and Malecki E J 2001 *Fixed and fluid: Stability and change in the geography of the Internet*. Paper presented at the annual meeting of the Association of American Geographers, New York.

Gorman S and McIntee A 2002 *Making sense of the urban spectrum: Agglomeration and location of wireless technologies*. Unpublished manuscript.

Gottmann J 1961 *Megalopolis* New York: The Twentieth Century Fund.

Graham S 1997 Cities in the real-time age: The paradigm challenge of telecommunications to the conception and planning of urban space *Environment and Planning A* **29**: 105–27.

Graham S 1998 The end of geography or the explosion of place? Conceptualizing space, place and information technology *Progress in Human Geography* **22**: 165–85.

Graham S 2001 The city as sociotechnical process: Networking mobilities and urban social inequalities *City* **5**: 339–49.

Graham S 2002 In a moment: On glocal mobilities and the terrorised city *City* **6** (forthcoming).

Graham S and Aurigi A 1997 Virtual cities, social polarization, and the crisis in urban public space *Journal of Urban Technology* **4**: 19–52.

REFERENCES

Graham S and Marvin S 1996 *Telecommunications and the City: Electronic Spaces, Urban Places* London: Routledge.

Graham S and Marvin S 2000 Urban planning and the technological future of cities. In J O Wheeler, Y Aoyama, and B Warf (eds) *Cities in the Telecommunications Age: The Fracturing of Geographies* New York: Routledge, 71–96.

Graham S and Marvin S 2001 *Splintering Urbanism: Networked Infrastructures, Technological Mobilities and the Urban Condition* London: Routledge.

Gregory D 1994 *Geographical Imaginations* Cambridge, MA: Blackwell.

Gregory D, Martin M and Smith G 1994 Introduction: Human geography, social change and social science. In D Gregory, R Martin, and G Smith, (eds) *Human Geography: Society, Space and Social Science* London: Macmillan, 78–112.

Grimes S 2000 Rural areas in the information society: Diminishing distance or increasing learning capacity? *Journal of Rural Studies* **16**: 13–21.

Grote M H, Harrschar-Ehrnborg S and Lo V 2002 Frankfurt's changing role in the European financial center system: A value chain approach. *Tijdschrift voor Economische en Sociale Geografie* **93** (forthcoming).

Grubesic T H and Murray A T 2002 Constructing the divide: Spatial disparities in broadband access *Papers in Regional Science* **81** (forthcoming).

Grubesic T H and O'Kelly M E 2002 Using points of presence to measure accessibility to the commercial Internet *The Professional Geographer* **54** (forthcoming).

Gupta A and Ferguson J 1992 Beyond 'culture': Space, identity, and the politics of difference. *Cultural Anthropology* **7**: 6–23.

Hackler D 2000 Industrial location in the information age: An analysis of information-technology-intensive industries. In J O Wheeler, Y Aoyama and B Warf (eds) *Cities in the Telecommunications Age: The Fracturing of Geographies* New York: Routledge, 200–18.

Halal W E 1993 The information technology revolution: Computer hardware, software, and services into the 21st century *Technological Forecasting and Social Change* **44**: 69–86.

Halavais A 2000 National borders on the world wide web *New Media and Society* **2**: 7–28.

Halbwachs M 1980 *The Collective Memory* F J Dulles and V Y Ditter (trans.) New York: Harper and Row.

Hall P 1997 Modelling the post-industrial city *Futures* **29**: 311–22.

Harasim L M 1993 Networlds: Networks as social space. In L M Harasim (ed.) *Global Networks: Computers and International Communication* Cambridge, MA: MIT Press, 16–34.

Hargittai E 1999 Weaving the Western web: Explaining differences in Internet connectivity among OECD countries *Telecommunications Policy* **23**: 701–18.

Harris R 1998 The Internet as a GPT: Factor market implications. In E Helpman (ed.) *General Purpose Technologies and Economic Growth* Cambridge, MA: MIT Press, 140–65.

Harvey D 1985 *The Urbanization of Capital* Oxford: Blackwell.

Harvey D 1989 *The Coming of Postmodernity* Oxford: Blackwell.

Harvey D 1993 From space to place and back again: Reflections on the condition of postmodernity. In J Bird, B Curtis, T Putnaman, G Robertson and L Tickner (eds) *Mapping the Futures: Local Cultures, Global Change* London: Routledge, 3–29.

Haug P 1991 Regional formation of high-technology service industries: The software industry in Washington State *Environment and Planning A* **23**: 869–84.

Helleiner E 1994 *States and the Reemergence of Global Finance* Ithaca: Cornell University Press.

Helpman E 1998 Introduction. In E Helpman (ed.) *General Purpose Technologies and Economic Growth* Cambridge, MA: MIT Press, 1–14.

Hepworth M E 1990 *Geography of the Information Economy* New York: Guilford.

Hewson M and Sinclair T J (eds) 1999 *Approaches to Global Governance Theory* Albany: State University of New York Press.

Heydebrand W 1999 Multimedia networks, globalization and strategies of innovation: The case of Silicon Alley. In H J Braczyk, G Fuchs and H G Wolf (eds) *Multimedia and Regional Economic Restructuring* London: Routledge, 49–80.

Hill R 1997 Electronic commerce, the World Wide Web, Minitel, and EDI *The Information Society* **13**: 33–41.

Hillis K 1998 On the margins: The invisibility of communications in geography *Progress in Human Geography* **22**: 543–66.

Hillner J 2000 Venture capitals. *Wired* **8.07**. http://www.wired.com/wired/archive/8.07/silicon.html

Hodgson G 1999 *Economics and Utopia: Why the Learning Economy is not the End of History* London: Routledge.

Holloway S I and Valentine G 2001 Placing cyberspace: Processes of Americanization in British children's use of the Internet *Area* **33**: 153–60.

Hopkins J 1994 *A mapping of cinematic places: Icons, ideology, and the power of (mis)representation*. In S C Aitken and L E Zonn (eds) *Place, Power, Situation and Spectacle: A Geography of Film* Lanham, MD: Roseman and Littlefield, 47–65.

Horrigan J H 2001 *Cities online: Urban development and the Internet* Pew Internet and American Life Project. http://www/pewinternet.org/reports/toc.asp?Report =50/

Host S 1991 The Norwegian newspaper system: Structure and development. In H Ronning and K Lundby (eds) *Media and Communication: Readings in Methodology, History and Culture* Oslo: Norwegian University Press, 281–301.

Hotz-Hart B 2000 Innovation networks, regions, and globalization. In G L Clark, M P Feldman and M S Gertler (eds) *The Oxford Handbook of Economic Geography* New York: Oxford, 432–50.

Howells J 2000 Knowledge, innovation and location. In J R Bryson, P W Daniels, N Henry and J Pollard (eds) *Knowledge, Space, Economy* London: Routledge, 50–62.

Hugill P J 1999 *Global Communications since 1844: Geopolitics and Technology* Baltimore: The Johns Hopkins University Press.

Huh W K and Kim H 2001 *Information flows on the Internet of Korea* Paper presented at the Digital Communities Conference, Chicago.

Ilchman W F 1970 New time in old clocks: productivity, development and comparative public administration. In D Waldo (ed.) *Temporal Dimensions of Development Administration* Durham, NC: Duke University Press, 135–78.

International Telecommunication Union (ITU) 1994 *Direction of Traffic: International Telephone Traffic 1994* Washington, D.C.: Telegeography.

International Telecommunication Union (ITU) 2001 Mobile on the eve of 3G *ITU Telecommunication Indicator Update* **April-June**: 1–3

REFERENCES

Internet.com. 2001 *The big picture geographics*. wysiwyg;//8/http://cyberatlas.inter-
net... raphics/article/0,,5911_732051,00.html

ISC (Internet Software Consortium) 2001 http://www.isc.org/ds/new-survey.html

Jaffe A B, Trajtenberg M and Henderson R 1993 Geographic localization of knowledge
spillover as evidenced by patent citations *The Quarterly Journal of Economics* **108**:
577–98.

Janelle D G 1968 Central place development in a time–space framework *The Professional Geographer* **20**: 5–10.

Janelle D G 1969 Spatial reorganization: A model and concept *Annals of the Association of American Geographers* **59**: 348–64.

Janelle D G 1991 Global interdependence and its consequences. In S D Brunn
and T R Leinbach (eds) *Collapsing Space and Time: Geographic Aspects of
Communication and Information* London: Harper Collins Academic, 49–81.

Janelle D G and Hodge D (eds) 2000 *Information, Place, and Cyberspace: Issues in
Accessibility* New York: Springer.

Jin D J and Stough R R 1998 Learning and learning capability in the Fordist and post-
Fordist age: An integrative framework *Environment and Planning A* **30**: 1255–78.

Johnson B and Lundvall B 2001 *Why all this fuss about codified and tacit knowledge?*
Paper presented at DRUID Winter Conference.

Katz P L 1988 *The Information Society: An International Perspective* New York:
Praeger.

Kellerman A 1985 The evolution of service economies: A geographical perspective
The Professional Geographer **37**: 133–43.

Kellerman A 1986 The diffusion of BITNET: A communications system for universities
Telecommunications Policy **10**: 88–92.

Kellerman A 1989 *Time, Space and Society: Geographical–Societal Perspectives* Dordrecht: Kluwer.

Kellerman A 1993a *Telecommunications and Geography* London: Belhaven.

Kellerman A 1993b *Society and Settlement: Jewish Land of Israel in the Twentieth
Century* Albany: State University of New York Press.

Kellerman A 1997 Fusions of information types, media, and operators, and continued American leadership in telecommunications *Telecommunications Policy* **21**:
553–64.

Kellerman A 1999a Leading nations in the adoption of communications media
1975–1995. *Urban Geography* **20**: 377–89.

Kellerman A 1999b Space and place in Internet information flows *NETCOM* **13**: 25–35.

Kellerman A 2000a Where does it happen? The location of the production and consumption of Web information *Journal of Urban Technology* **7**: 45–61.

Kellerman A 2000b Phases in the rise of information society. *Info* **2**: 537–41.

Kellerman A 2002a Conditions for the development of high-tech industry: The case
of Israel. *Tijdschrift voor Economische en Sociale Geografie* **93**: 270–86.

Kellerman A 2002b The global leadership of New York and Los Angeles in information
production *Journal of Information Technology* **9**: 21–35.

Kenney M and Curry J 2001 Beyond transaction costs: E-commerce and the power of
the Internet dataspace. In T R Leinbach and S D Brunn (eds) *Worlds of E-commerce:
Economic, Geographical and Social Dimensions* Chichester: Wiley, 45–66.

Keohane R O and J S Nye Jr 1998 Power and interdependence in the information age *Foreign Affairs* **77**: 81–94.

King A 1990 Architecture, capital and the globalization of culture *Theory, Culture and Society* 7: 397–411.

Kipnis B A 2001 Tel Aviv, Israel – A world city in evolution: Urban development in a deadend of the global economy *Globalization and World Cities Study Group and Networks Research Bulletin* **57**, http://www.lboro.ac.uk/gawc/rb/rb57.html.

Kipnis B A 2002 The impact of globalization on the boom and crisis of Israel's high-tech industry *Globalization and World Cities Study Group and Networks Research Bulletin* **72**, http://www.lboro.ac.uk/gawc/rb/rb72.html.

Kirsch S 1995 The incredible shrinking world? Technology and the production of space *Environment and Planning D: Society and Space* **13**: 529–55.

Kitchin R 1998 *Cyberspace: The World in the Wires* Chichester: Wiley.

KNIC (Korea Network Information Center) 2002 *A survey on the Internet users and use behavior* (Korean) Received through W K Huh.

Knox P L 1995 World cities and the organization of global space. In R J Johnston, P J Taylor and M J Watts (eds) *Geographies of Global Change: Remapping the World in the Late Twentieth Century* Oxford: Blackwell, 232–47.

Kobrin S 1997 Electronic cash and the end of national markets *Foreign Policy* (summer) **107**: 65–77.

Kopomaa T 2000 *The City in Your Pocket: Birth of the Mobile Information Society* Helsinki: Gaudeamus.

Kornbluh K 2001 The broadband economy *The New York Times*, December 10, http://www.nytimes.com/2001/12/10/opinion/10KORN.html?todaysheadlines/

Kotkin J 2000 *The New Geography: How the Digital Revolution is Reshaping the American Landscape* New York: Random House.

Kwan M P 2001a Cyberspatial cognition and individual access to information: The behavioral foundation of cybergeography *Environment and Planning B: Planning and Design* **28**: 21–37.

Kwan M P 2001b *Time, space, information technologies and urban geographies* Paper presented at the Digital Communities Conference, Chicago.

Lakshmanan T R, Anderson D E, Chatterjee L and Sasaki K 2000 Three global cities: New York, London and Tokyo. In Å E Andersson and D E Andersson (eds) *Gateways to the Global Economy* Northampton, MA: Edward Elgar, 49–81.

Langdale J 2001 Global electronic spaces: Singapore's role in the foreign exchange market in the Asia–Pacific region. In T R Leinbach and S D Brunn (eds) *Worlds of E-commerce: Economic, Geographical and Social Dimensions* Chichester: Wiley, 203–21.

Lash S and Urry J 1994 *Economies of Signs and Space* London: Sage.

Latour B 1987 *Science in Action: How to Follow Scientists and Engineers through Society* Cambridge, Ma: Harvard University Press.

Laulajainen R 1998 *Financial Geography: A Banker's View* Gothenburg: Gothenburg School of Economics and Commercial Law.

Lawrence S and Giles C L 1999 Accessibility of information on the web *Nature* **400**: 107–9.

Lefebvre H 1991 *The Production of Space* D Nicholson-Smith, (trans.) Oxford: Basil Blackwell.

227

REFERENCES

Lefebvre H 1974 *La Production de l'espace*. Paris, Anthropos.

Leinbach T R 2001 Emergence of the digital economy and e-commerce. In T R Leinbach and S D Brunn (eds) *Worlds of E-commerce: Economic, Geographical and Social Dimensions* Chichester: Wiley, 3–26.

Leinbach T R and Brunn S D 2001 E-commerce: Definitions, dimensions and constraints. In T R Leinbach and S D Brunn (eds) *Worlds of E-commerce: Economic, Geographical and Social Dimensions* Chichester: Wiley, xi–xiii.

Lemos A 1996 The labyrinth of Minitel. In R Shields (ed.) *Cultures of Internet: Virtual Spaces, Real Histories, Living Bodies* London: Sage, 33–48.

Lessig L 2002 Who's holding back broadband? *The Washington Post* January 8. http://www.washingtonpost.com/ac2/wp-dyn?pagename=article&node=&contentld =A1

Li F 1995 *The Geography of Business Information* Chichester: Wiley.

Li F, Whalley J and Williams H 2001 Between physical and electronic spaces: The implications for organizations in the networked economy *Environment and Planning A* **33**: 699–716.

Locksley G 1992 The information business. In K Robins (ed.) *Understanding Information Business, Technology and Geography* London: Belhaven.

Lou B K 2001 *The rise of digital communities in Asia: A case study of the People's Republic of China* Paper presented at the Digital Communities Conference, Chicago.

Luger M and Goldstein H 1991 *Technology in the Garden: Research Parks and Regional Economic Development* Chapel Hill, NC: University of North Carolina Press.

Luke T W and Tuathail G O 1998 Global flowmations, local fundamentalisms, and fast geopolitics. In A Herod, G O Tuathail and S M Roberts (eds) *An Unruly World? Globalization, Governance and Geography* London: Routledge, 72–94.

Lundvall B A and Maskell P 2000 Nation states and economic development: From national systems of production to national systems of knowledge creation and learning. In G L Clark, M P Feldman and M S Gertler (eds) *The Oxford Handbook of Economic Geography* New York: Oxford University Press, 353–72.

Lycos 1999, 2001 *The Lycos 50 daily report* http://50.lycos.com/.

Lyman P and Varian H R 2000 *How Much Information?* School of Information Management and Systems, University of California, Berkeley. http://www.sims.berkeley.edu/research/projects/how-much-info/.

Lyon D 1988 *The Information Society: Issues and Illusions* Cambridge: Polity Press.

Lyon D 1995 The roots of the information society idea. In N Heap, R Thomas, G Einon, R Mason and H Mackay (eds) *Information Technology and Society: A Reader* 54–73.

Ma'ariv 1999 October, 3 (Hebrew).

Machlup F 1962 *The Production and Distribution of Knowledge in the United States* Princeton, NJ: Princeton University Press.

Machlup F 1980 *Knowledge and Knowledge Production* Princeton: Princeton University Press.

Machlup F 1983 Semiotic quirks in studies of information. In F Machlup and U Mansfield (eds) *The Study of Information: Interdisciplinary Messages* New York: Wiley 641–71.

Makridakis S 1995 The forthcoming information revolution *Futures* **27**: 799–821.

Malecki E J 2000a Creating and sustaining competitiveness: Local knowledge and economic geography. In J R Bryson, P W Daniels, N Henry and J Pollard (eds) *Knowledge, Space, Economy* London: Routledge, 103–19.

Malecki E J 2000b *The Internet: Its economic geography and policy implications*. Paper presented at the Global Economic Geography Conference, Singapore.

Malecki E J 2002 The economic geography of the Internet's infrastructure *Economic Geography* **78** (forthcoming).

Malecki E J and Gorman S P 2001 Maybe the death of distance, but not the end of geography: The Internet as a network. In T R Leinbach and S D Brunn (eds) *Worlds of E-commerce: Economic, Geographical and Social Dimensions* Chichester: Wiley, 87–105.

Malecki E J and McIntee A 2001 *Making the connections: Interconnecting the networks of the Internet* Paper presented at the Digital Communities conference, Chicago.

Mansfield E J 1991 Academic research and industrial innovation *Research Policy* **20**: 1–12.

Mapping a Future for Digital Connections: A Study of the Digital Divide in San Diego County http://www.sdrta.org

Marable L 2001 Broadband – it's a city thing http://www.idg.net/crd_idgsearch_540 830.html?sc=

Markoff J and Richtel M 2002 Signs of rebound appear in the high-tech heartland *New York Times* January 14.

Markusen A 1996 Sticky places in a slippery space: A typology of industrial districts *Economic Geography* **72**: 293–313.

Markusen A, Chapple K, Schrock G, Yamamoto D and Yu P 2001 *High-Tech and I-Tech: How Metros Rank and Specialize* Minneapolis, MN: Project on Regional and Industrial Economics, Humphrey Institute of Public Affairs, University of Minnesota.

Martin W E 2001 Down on the server farm *GovTech*. http://www.govtech.net/

Martin W J 1988 *The Information Society* London: Aslib.

Martin W J 1995 *The Global Information Society* Aldershot: Aslib Gower.

Marvin C 1988 *When Old Technologies Were New: Thinking about Electric Communication in the Late Nineteenth Century* New York: Oxford University Press.

Maskell P 1999 Social capital and regional development *North* **10**: 9–13.

Maskell P 2000 Social capital, innovation and competitiveness. In S Baron, J Field and T Schuller (eds) *Social Capital: Critical Perspectives* Oxford: Oxford University Press, 111–23.

Maskell P, Eskelinen H, Hannibalsson I, Malmberg A and Vatne E 1998 *Competitiveness, Learning and Regional Development* London: Routledge.

Maskell P and Malmberg H 1999a The competitiveness of firms and regions: 'Ubiquitification' and the importance of localized learning *European Urban and Regional Studies* **6**: 9–25.

Maskell P and Malmberg H 1999b Localized learning and industrial competitiveness *Cambridge Journal of Economics* **23**: 167–85.

Massey D 1992 Politics and space/time *New Left Review* **196**: 65–84.

Massey D 1994 *Space, Place and Gender* Cambridge: Polity Press.

Masuda Y 1980 *The Information Society as Post-Industrial Society* Washington, DC: World Future Society.

REFERENCES

Media Matrix 1999, 2001 *Global top 50 Web and digital media properties* http://www.relevantknowledge.com/.

Merrifield A 1993 Place and space: A Lefebvrian reconciliation *Transactions of the British Institute of Geographers* **18**: 516–31.

Meyrowitz J 1985 *No Sense of Place: The Impact of Electronic Media on Social Behavior* New York: Oxford University Press.

Miles I and Robins K 1992 Making sense of information. In Robins K (ed.) *Understanding Information Business, Technology and Geography* London: Belhaven, 1–26.

Mitchell W M 1995 *City of Bits: Space, Place, and the Infobahn* Cambridge, MA: MIT Press.

Mitchelson R L 1999 Spatial aspects of technology transfer: Evidence from Kentucky. *Southeastern Geographer* **39**: 206–19.

Mitchelson R L and Wheeler J O 1994 The flow of information in a global economy: The role of the American urban system in 1990 *Annals of the Association of American Geographers* **84**: 87–107.

Moore D 2001 The spread of the Code-Red worm (CRv2) http://www.caida.org/analysis/security/code-red/

Morgan K 1992 Digital highways: The new telecommunications era *Geoforum* **23**: 317–32.

Morley D and Robins K 1995 *Spaces of Identity: Global Media, Electronic Landscapes and Cultural Boundaries* London: Routledge.

Morris M and Ogan C 1994 The Internet as mass medium *Journal of Computer Mediated Communication* **1**. http://www.ascuse.org/jcmc/vol1/issue4/morris.html.

Mosco V 1996 *The Political Economy of Communication: Rethinking and Renewal* London: Sage.

Moss M L 1986 Telecommunications systems and large world cities: A case study of New York. In A D Lipman, A D Sugerman and R F Cushman (eds) *Teleports and the Intelligent Cities* Homewood, IL: Dow Jones-Irwin, 378–97.

Moss M L 1996 Telecommunications policy and cities http://www.nyu.edu/urban/research/telecom/telecom.html.

Moss M 2000 Why New York will flourish in the 21st century *New York Observer*, January 10th.

Moss M L and Townsend A 1997 Tracking the net: Using domain names to measure the growth of the Internet in U.S. cities *Journal of Urban Technology* **4**: 47–60.

Moss M L and Townsend A 1998 Spatial analysis of the Internet in U.S. cities and states http://urban.nyu.edu/research/newcastle/.

Moss M L and Townsend A M 2000 The Internet backbone and the American metropolis *The Information Society* **16**: 35–47.

Mulgan G 1989 New times: A tale of new cities *Marxism Today* March, 18–24.

Negroponte N 1995 *Being Digital* New York: Alfred A. Knopf

NetValue 2001 *Korea leads world in broadband usage* http://www.netvalue.com/corp/presse/cp0028.htm

Nielsen Media Research 2000 *2000 Report on Television* New York.

Nielsen/NetRatings 2000 *Internet penetration and usage for 20 local markets.* http://www.nielsen-netratings.com/press_releases/pr_000412.htm

Nijman J 2000 The paradigmatic city *Annals of the Association of American Geographers* **90**: 135–45.

Noam E M 1997 Beyond liberalization: From the network of networks to the system of systems http://www.ctr.columbia.edu/vii/papers/citi698.htm.

Nolan R L 2000 Information technology management since 1960. In A D Chandler Jr and J W Cortada (eds) *A Nation Transformed by Information: How Information Has Shaped the United States from Colonial Times to the Present* New York: Oxford University Press, 217–56.

NTIA (National Telecommunications and Information Administration) 2000 *Telecom Glossary 2000* http://www.its.bldrdoc.gov/projects/tlglossary2000/_data_switching_exchange.html.

NUA 2001 *US ecommerce 1998–2003.* http://www.nua.ie/surveys/analysis...arts/comparisons/ecommerce_us.html

Nye J S Jr and Owens W A 1996 America's information edge *Foreign Affairs* **75**: 20–36.

OECD 1997 *Webcasting and convergence: Policy implications* http://www.oecd.org/dsti/sti/it/cm/prod/e_97-221.htm.

OECD 2000a *Knowledge Management in the Learning Society* Paris: Center for Education and Research and Innovation.

OECD 2000b *OECD Information Technology Outlook 2000: ICTs, E-commerce and the Information Economy* Paris: Organization for Economic Co-operation and Development.

OECD 2001 *The Development of Broadband Access in OECD Countries* Working Party on Telecommunication and Information Services Policies http://www.oecd.org/pdf/m00020000/m00020255.pdf

Ogden M R 1994 Politics in a parallel universe *Futures* **26**: 713–29.

Ohmae K 1993 The rise of the region state *Foreign Affairs* **72**: 78–87.

Ó hUallacháin B 1999 Patent places: Size matters *Journal of Regional Science* **39**: 613–36.

O'Kelly M E and Grubesic T H 2002 Backbone topology, access and the commercial Internet, 1997–2000 *Environment and Planning B* **29** (forthcoming).

O'Neill H O and Moss M L 1991 *Reinventing New York: Competing in the Next Century's Global Economy* New York: Urban Research Center, New York University.

Ostbye H 1997 Norwegian media in the 1990s: Internationalization, concentration and commercialization. In U Carlsson and E Harrie (eds) *Media Trends 1997: In Denmark, Finland, Iceland, Norway and Sweden* Goteborg: NORDICOM Goteborg University, 67–86.

O'Toole L 1998 *Pornocopia: Porn, Sex, Technology and Desire* London: Serpent's Tail.

Palfini J 2001 Why it's getting easier to get your teenager off the phones *The Industry Standard* wysiwyg://2/file:/c|/WINDOWS/Tempora...Get Your Teenager Off the Phone.html

Palmer J W 1997 Electronic commerce in retailing: Differences across retail formats *The Information Society* **13**: 75–92.

Paltridge S 1999 Mining and mapping web content *Info* **1**: 327–42.

Paltridge S 2001 Local access pricing and the international digital divide *On the Internet* http://www.isoc.org/oti/articles/1000/paltridge.html

Park T M 2000 *Analysis – factors leading to sharp increase Internet users in Korea* Korea Network Information Center (KNIC), Analysis Report.

Parker E B 2000 Closing the digital divide in rural America *Telecommunications Policy* **24**: 281–90.

REFERENCES

Pastore M 2001 *U.S. broadband users concentrated in cities. Internet.com.* http: //cyberatlas.internet.com/markets/broadband/article/0,,10099_766351,00.html

Patel P and Pavitt K 1991 Large firms in the production of the world's technology: An important case of 'non-globalization' *Journal of International Business Studies* **22**: 1–21.

Pavlik J V 1999 Contents and economics in the multimedia industry: The case of New York's Silicon Alley. In H J Braczyk, G Fuchs and H G Wolf (eds) *Multimedia and Regional Economic Restructuring* London: Routledge, 81–96.

Pelletiere D and Rodrigo G C 2001 An empirical investigation of the digital divide in the United States http://www.delft2001.tude1ft.nl/paper%20files/paper1102.doc

Pierce S 2001 *Location, location, location* http://www.research.microsoft.com/re-search/sn/location.asp

Population Reference Bureau 2001 http://www.prb.org/Content/NavigationMenu/Other_Reports/2000–2002/sheet1.html

Porat M 1977 *The Information Economy: Definition and Measurement* Washington, DC: US Department of Commerce, Office of Telecommunications, Publication 77–12 (1).

Postman N 1999 *Building a Bridge to the Eighteenth Century* New York: Alfred A. Knopf.

Pred A 1977 *City Systems in Advanced Economies* London: Hutchinson.

Pred A 1984 Place as historically contingent process: Structuration theory and the time-geography of becoming places *Annals of the Association of American Geographers* **74**: 279–97.

Pred A and Watts M J 1992 *Reworking Modernity: Capitalisms and Symbolic Discontent* New Brunswick, NJ: Rutgers University Press.

Quay R 2001 Bridging the digital divide *Planning Magazine* http://www.planning.org/pubs/plng01/july011.htm

Reed H 1981 *The Preeminence of International Financial Centers* New York: Praeger.

Regan P M 2001 'Dry counties' in cyberspace: Governance and enforcement without geographic boundaries. In T R Leinbach and S D Brunn (eds) *Worlds of E-commerce: Economic, Geographical and Social Dimensions* Chichester: Wiley, 257–76.

Relph E 1976 *Place and Placelessness* London: Pion.

Rheingold H 1993a *The Virtual Community: Homesteading on the Electronic Frontier* Reading: Addison-Wesley.

Rheingold H 1993b A slice of life in my virtual community. In L M Harasim (ed.) *Global Networks: Computers and International Communication* Cambridge, MA: MIT Press, 57–82.

Richardson R and Gillespie A 2000 The economic development of peripheral rural areas in the information age. In M I Wilson and K E Corey (eds) *Information Tectonics: Space, Place, and Technology in an Electronic Age* Chichester: Wiley, 199–217.

Richtel M 2001 A city takes a breath after the dot-com crash *New York Times*, July 24.

Roberts J 2000 *The drive to codify: Implications for the knowledge-based economy* Paper presented at the eighth International Joseph A. Schumpeter Society Conference, University of Manchester.

Roberts S 1994 Fictitious capital, fictitious spaces: The geography of offshore financial flows. In S Corbridge, N Thrift and R Martin (eds) *Money, Power and Space* Oxford: Blackwell, 91–115.

Robertson R 1995 'Glocalisation': Time-space and homogeneity-heterogeneity. In M Featherstone S Lash and R Robertson (eds) *Global Modernities* London: Sage, 25–44.

Robins K (ed.) 1992 *Understanding Information Business, Technology and Geography* London: Belhaven.

Robins K and Cornford J 1994 Local and regional broadcasting in the new media order. In A Amin and N Thrift (eds) *Globalization, Institutions, and Regional Development in Europe* Oxford: Oxford University Press, 217–38.

Romero S 2001 *Wireless cooly received in U.S.* http://nytimes.com/2001/01/29/technology/29WIRE.html

Roos J P 1994 A post-modern mystery: Why do Finns, 'silent in two languages', have the highest density of mobiles in the world? *Intermedia* **22**: 24–8.

Rosencrance R 1999 *The Rise of the Virtual State: Wealth and Power in the Coming Century* New York: Basic Books.

Roszak T 1991 *The Cult of Information: A neo-Luddite treatise on high-tech, artificial intelligence and the true art of thinking*, 2nd ed. Berkeley: University of California Press.

Sack R 1980 *Conceptions of Space in Social Thought* London: Macmillan.

Sassen S 1991 *The Global City: New York, London, Tokyo* Princeton: Princeton University Press.

Sassen S 1994 *Cities in a World Economy* Thousand Oaks, CA: Pine Forge Press.

Sassen S 1997 The new centrality: The impact of telematics and globalization. In P Droege (ed.) *Intelligent Environments: Spatial Aspects of the Information Revolution* Amsterdam: Elsevier, 19–27.

Sassen S 1999 Global financial centers *Foreign Affairs* **78**: 75–87.

Sassen S 2001a Cities in the global economy. In R Paddison (ed.) *Handbook of Urban Studies* London: Sage, 256–72.

Sassen S 2001b Locating cities on global circuits *Globalization and World Cities Study Group and Networks Research Bulletin* **46**, http://www.lboro.ac.uk/gawc/rb/rb46.html.

Saxenian A 1994 *Regional Advantage: Culture and Competition in Silicon Valley and Route 128* Cambridge, MA: Harvard University Press.

Scarborough 2000 *Almost half of Internet users are buying products or services online* http://www.scarborough.com/scarb2002/press/pr_cyber_announce.htm

Schement I R 1989 The origins of the information society in the United States: Competing visions. In J L Salvaggio (ed.) *The Information Society: Economic, Social, and Structural Issues* Hillsdale, NJ: Lawrence Erlbaum Associates, 29–50.

Schiller H I 1981 *Who Knows: Information in the Age of the Fortune 500* Norwood, NJ: Ablex.

Schivelbusch W 1978 Railroad space and railroad time *New German Critique* **14**: 31–40.

Schrag Z M 1994 Navigating cyberspace – maps and agents: Different uses of computer networks call for different interfaces. In G C Staple (ed.) *Telegeography 1994: Global Telecommunications Traffic* Washington, DC: Telegeography, Inc., 44–52.

Schrag Z M 1996 The achilles heel of Internet telephony. In G Staple (ed.) *Telegeography 1996/97* Washington, DC: Telegeography, 37–40.

SCMP (South China Morning Post) 2002 Global surge in system attacks, January 30.

REFERENCES

Scott A J 1993 *Technopoles: High-Technology Industry and Regional Development in Southern California* Los Angeles: University of California Press.

Scott A J 1995 *From Silicon Valley to Hollywood: Growth and development of the multimedia industry in California* Working Paper 13, The Lewis Center for Regional Policy Studies, University of California, Los Angeles.

Scott A J 2001 Capitalism, cities, and the production of symbolic forms *Transactions of the Institute of British Geographers* **26**: 11–23.

Shamai S and Kellerman A 1985 Conceptual and experimental aspects of regional awareness: an Israeli case study *Tijdschrift voor Economische en Sociale Geografie* **76**: 88–99.

Shefer A and Frenkel A 1998 Local milieu and innovations: Some empirical results *The Annals of Regional Science* **32**: 185–200.

Sheppard E, Couclelis H, Graham S, Harrington J W and Onsrud H 1999 Geographies of the information society *International Journal of Geographical Information Science* **13**: 797–823.

Shields R 1996 Introduction: Virtual spaces, real histories and living bodies. In R Shields, (ed.) *Cultures of Internet: Virtual Spaces, Real Histories, Living Bodies* London: Sage, 1–10.

Shiva V 1993 The greening of the global reach. In W Sachs (ed.) *Global Ecology: A New Arena of Political Conflict* London: Zed Books, 149–56.

Short J R and Kim Y H 1999 *Globalization and the City* Essex: Addison Wesley Longman.

Siemens 1975, 1986 *International Telecom Statistics 1975, 1986* Munich: Siemens.

Silverstone R 1993 Time, information and communication technologies and the household *Time and Society* **2**: 283–311.

Simonsen K 1996 What kind of space in what kind of social theory? *Progress in Human Geography* **20**: 494–512.

Simpson J A and Weiner E S C 1989 *The Oxford English Dictionary* 2nd ed. Oxford: Clarendon.

Smith D F Jr and Florida R 2000 Venture capital's role in regional innovation systems: Historical perspective and recent evidence. In Z J Acs (ed.) *Regional Innovation, Knowledge and Global Change* London: Pinter, 205–27.

Smith G V and Parr R L 2000 *Valuation of Intellectual Property and Intangible Assets.* 3rd ed. New York: Wiley.

Smith M P 2000 *Transnational Urbanism: Locating Globalization* Oxford: Blackwell.

Smith N 1993 Homeless/global: Scaling places. In J Bird, B Curtis, T Putnam, G Robertson and L Tickner (eds) *Mapping the Futures: Local Cultures, Global Change* London: Routledge, 87–119.

Soja E W 1989 *Postmodern Geographies: The Reassertion of Space in Critical Social Theory* London: Verso.

Soja E W 2000 *Postmetropolis: Critical Studies of Cities and Regions* Oxford: Blackwell.

Sölvell Ö and Birkinshaw J 2000 Multinational enterprises and the knowledge economy: Leveraging global practices. In J H Dunning (ed.) *Regions, Globalization, and the Knowledge-Based Economy* New York: Oxford, 82–106.

Sproul L S 2000 Computers in U.S. households since 1977. In A D Chandler Jr and J W Cortada (eds) *A Nation Transformed by Information: How Information has*

Shaped the United States from Colonial Times to the Present New York: Oxford University Press, 257–80.

Steinfeld C and Salvaggio J L 1989 Toward a definition of information society. In J L Salvaggio (ed.) *The Information Society: Economic, Social, and Structural Issues* Hillsdale, NJ: Lawrence Erlbaum Associates, 1–14.

Sternberg R and Arndt O 2001 The firm or the region: What determines the innovation behaviour of European firms? *Economic Geography* **77**: 364–82.

Stewart T A 1997 *Intellectual Capital* London: Nicholas Bradley.

Storper M 1997 *The Regional World: Territorial Development in a Global Economy* New York: Guilford.

Storper M 2000a Globalization, localization, and trade. In G L Clark, M P Feldman and M S Gertler (eds) *The Oxford Handbook of Economic Geography* New York: Oxford University Press, 146–65.

Storper M 2000b Globalization and knowledge flows: An industrial geographer's perspective. In J H Dunning 2000 *Regions, Globalization, and the Knowledge-Based Economy* New York: Oxford, 42–62.

Storper M and Christopherson S 1987 Flexible specialization and regional industrial agglomerations: The case of the U.S. motion picture industry *Annals of the Association of American Geographers* **77**: 104–17.

Swartz J 2001 Houston tackles 'digital divide' with free e-mail *USA Today*, August 20.

Swyngedouw E A 1992a Territorial organization and the space/technology nexus *Transactions of the British Institute of Geographers* **17**: 417–33.

Swyngedouw E A 1992b The mammon quest 'glocalisation', interspatial competition and the monetary order: The construction of new scales. In N Dunford and G Kafkalas (eds) *Cities and Regions in the New Europe: The Global–Local Interplay and Spatial Development Strategies* London: Belhaven, 39–67.

Swyngedouw E A 1993 Communication, mobility and the struggle for power over space. In G Giannopoulos and A Gillespie (eds) *Transport and Communications in the New Europe* London: Belhaven, 305–25.

Swyngedouw E 1997 Neither global nor local: 'Glocalization' and the politics of scale. In K R Cox, (ed.) *Spaces of Globalization: Reasserting the Power of the Local* New York: Guilford, 137–66.

Telecom Finland 1997 – pioneer in the opening of the telecommunications market for competition October 1997 *Globes* (Business Focus) 8–9 (Hebrew).

TeleGeography 2000: *Global Telecommunications Traffic Statistics and Commentary* 1999 Washington, DC: TeleGeography.

TeleGeography 2001: *Global Telecommunications Traffic Statistics and Commentary* 2000 Washington, DC: TeleGeography.

TeleGeography 2001 *What's new?* http://www.telegeography.com/Whatsnew/pg02

TeleGeography 2002 *Global Telecommunications Traffic Statistics and Commentary* 2001 Washington, DC: TeleGeography.

Thomson Financial Investor Relations 1999 *1999 International Target Cities Report* New York.

Thrift N 1985 Files and germs: A geography of knowledge. In D Gregory and J Urry (eds) *Social Relations and Spatial Structures* London: Macmillan, 366–403.

REFERENCES

Thrift N 1994a On the social and cultural determinants of international financial centers: The case of the city of London. In S Corbridge, R L Martin and N Thrift (eds) *Money, Power and Space* Oxford: Blackwell, 327–55.

Thrift N 1994b Taking aim at the heart of the region. In D R Gregory, R Martin and G Smith (eds) *Human Geography: Society, Space and Social Science* London: Macmillan, 200–31.

Thrift N 1995 A hyperactive world. In R J Johnston, P J Taylor and M J Watts (eds) *Geographies of Global Change: Remapping the World in the Late Twentieth Century* Oxford: Blackwell, 18–35.

Thrift N 1996 *Spatial Formations* London: Sage.

Thu Nguyen D and Alexander J 1996 The coming of cyberspacetime and the end of the polity. In R Shields (ed.) *Cultures of Internet: Virtual Spaces, Real Histories, Living Bodies* London: Sage, 99–124.

Tivers J 1996 *Landscapes of computer games*. Paper presented at the annual conference of the Institute of British Geographers.

Top U.S. Book Markets 2000 http://www.booknotes.org/market.htm.

Touraine A 1974 *The Postindustrial Society: Tomorrow's Social History: Classes, conflicts and Culture in the Programmed Society* L F X Mayhew (trans.) London: Wildwood House.

Townsend A 2000a Geography lessons for your server *TELECOM-CITIES Research Update* **1**:1–3, http://www.informationcity.org/telecom-cities.

Townsend A 2000b Life in the real-time city: Mobile telephones and urban metabolism *Journal of Urban Technology* **7**: 85–104.

Townsend A 2001a The Internet and the rise of the new network cities, 1969–1999 *Environment and Planning B: Planning and Design* **28**: 39–58.

Townsend A 2001b Network cities and the global structure of the Internet *American Behavioral Scientist* **44**: 1697–716.

Tseng K F and Litman B 1998 The impact of the Telecommunications Act 1996 on the merger of RBOCs and MSOs: Case study: The merger of US West and Continental Cablevision. *Journal of Media Economics* **11**: 47–64.

Tuan Y F 1991 Language and the making of place: A narrative-descriptive approach *Annals of the Association of American Geographers* **81**: 684–96.

Tuathail O and Luke T W 1994 Present at the (dis)integration: Deterritorialization and reterritorialization in the new wor(l)d order *Annals of the Association of American Geographers* **84**: 381–98.

TV Basics 2000 http://www.tvb.org/tvfacts/tvbasics/basics21.html.

Tyner J A 2000 Global cities and circuits of global labor: The case of Manila, Philippines. *Professional Geographer* **52**: 61–74.

Uimonen T 2001 *Net usage in Asia catching up with U.S.* http://www.idg.com.hk/cw/readstory.asp?aid =20010110009

Ullman E 1974 Space and/or time: opportunity for substitution and prediction *Transactions of the British Institute of Geographers* **63**: 135–39.

United Nations 1978, 1990, 1996 *Statistical Yearbook 1977;1987; 1994* New York: United Nations.

United Nations 1999 *The Human Development Report 1999* http://www.undp.org/hdro/Chapter2.pdf.

University of Texas 2001 *Measuring the Internet Economy* http://www.internetindica-tors.com

Unwin T 2000 A waste of Space? Towards a critique of the social production of space *Transactions of the Institute of British Geographers* **25**: 11–29.

Warf B 2001 Segueways into cyberspace: Multiple geographies of the digital divide *Environment and Planning B: Planning and Design* **28**: 3–19.

Warf B and Purcell D 2001 The currency of currency: Speed, sovereignty, and electronic finance. In T R Leinbach and S D Brunn (eds) *Worlds of E-commerce: Economic, Geographical and Social Dimensions* Chichester: Wiley, 223–40.

The Washington Post 2002 Government sites draw web traffic, January 9.

Webster F 1994 What information society? *The Information Society* **10**: 1–23.

Weibull L and Anshelm M 1992 Indications of Change: Developments in Swedish Media 1980–1990 *Gazette* **49**: 41–73.

Wellman B 2001 Computer networks as social networks *Science* **293**: 2031–4.

Wellman B and Gulia M 1999 Virtual communities as communities: Net surfers don't ride alone. In M A Smith and P Kollock (eds) *Communities in Cyberspace* London: Routledge, 167–94.

Westland J C and Clark T H K 1999 *Global Electronic Commerce: Theory and Case Studies* Cambridge, MA: The MIT Press.

Wheeler D C and O'Kelly M E 1999 Network topology and city accessibility of the commercial Internet *The Professional Geographer* **51**: 327–9.

Wigand R T 1997 Electronic commerce: Definition, theory, and context *The Information Society* **13**: 1–16.

Wilson M I 2001a Location, location, location: The geography of the dot com problem. *Environment and Planning A* **28**: 59–71.

Wilson M I 2001b Dot com development: Are IT lines better than tractors? In T R Leinbach and S D Brunn (eds) *Worlds of E-commerce: Economic, Geographical and Social Dimensions* Chichester: Wiley, 277–92.

Wilson M 2002 *Chips, bits, and the law: An economic geography of Internet gambling.* Paper presented at the Annual Meeting of the Association of American Geographers (AAG), Los Angeles.

Wilson M I, Corey K E, Mickens C and Pratt Mickens H 2001 *Death of distance/rise of place: The impact of the Internet on locality and spatial organization.* Paper presented at the 11th Annual Internet Society Conference, Stockholm http://www.isoc.org/inet2001/CD_proceedings/U128/INET2001-U128.htm.

The World Bank 1997 *World Development Indicators 1997* Washington, D.C.: The World Bank, 1997.

The World Bank 2001a *World Development Indicators 2001* http://devdata.worldbank.org

The World Bank 2001b *Global Information Infrastructure: Executive Summary* http://www.infodev.org/projects/375/fin375.htm

The WorldPaper 2001 Information society index http://www.worldpaper.com/ISI/Intro%20januaryfebruary2001.html.

Yeoh B S A 1999 Global/globalizing cities. *Progress in Human Geography* **23**: 607–16.

Zook M A 1998 *The web of consumption: The spatial organization of the Internet industry in the United States* http://socrates.berkeley.edu/~zook/pubs/acsp1998.html.

REFERENCES

Zook M A 1999 *Old hierarchies or new networks of centrality? – understanding the global geography of the Internet content market* Paper presented at the conference on Cities in the Global Information Society, Newcastle-Upon-Tyne.

Zook M A 2000 The web of production: The economic geography of commercial Internet content production in the United States *Environment and Planning A* **32**: 411–26.

Zook M A 2001a http://socrates.berkeley.edu/~zook/domain_names/.

Zook M A 2001b Old hierarchies or new networks of centrality? – understanding the global geography of the Internet content market *American Behavioral Scientist* **44**: 1679–96.

Zook M A 2001c *Peripheral nodes in the space of flows: The geography of the Internet adult industry* Paper presented at the Digital Communities Conference, Chicago.

Zook M 2001d Connected is a matter of geography *netWorker* **5**: 13–17.

Zook M A 2002 Grounded capital: Venture financing and the geography of the Internet industry, 1994–2000 *Journal of Economic Geography* **2** (forthcoming).

GLOSSARY OF
ABBREVIATIONS

APNIC	Asia-Pacific Network Information Center
ARIN	American Registry for Internet Numbers
ARPANET	Advanced Research Project Network
B2B	Business to business
B2C	Business to customers
BITNET	Because It's Time Network
CATV	Cable television
CBD	Central business district
CDMA	Code division multiple access
CIX	Commercial Internet exchange
CO	Central switching office
CONE	.com; .org; .net; .edu
DNS	Domain name system
DSE	Data switching exchange
DSL	Digital subscriber line
EARN	European Academic and Research Network
EDI	Electronic data interchange
EMS	Enhanced messaging service
FCC	Federal Communication Commission
FDI	Foreign direct investment
FIRE	Finance, insurance, and real estate
FTP	File transfer protocol
FX	Foreign exchange
GIS	Geographical information systems
GPS	Global positioning system
GPT	General purpose technology
GSM	Global system for mobile communications
HTML	Hypertext markup language

GLOSSARY OF ABBREVIATIONS

HTTP	Hypertext transfer protocol
IANA	Internet assigned number authority
ICT	Information and communications technology
IMF	International Monetary Fund
IP	Internet protocol
ISBN	International standard book number
ISDN	Integrated services digital network
ISP	Internet service provider
IT	Information technology
IX	Internet exchange
kbps	Kilobytes per second
LAN	Local area network
LBS	Location-based service
LIR	Local Internet registries
MAE	Metropolitan area exchange
mbps	Million bits per second
MMS	Multimedia messaging
MNC	Multi-national corporation
MOO	MUD (Multi-user dungeon) object oriented
NAP	Network access point
NASDAQ	National Association of Securities Dealers Automated Quotation
NSFNET	National Science Foundation Network
P2P	Peer-to-peer
POP	Point of presence
R&D	Research and development
RIPE	Réseaux IP Européens
SMA	Standard metropolitan area
SMS	Short message service
SWIFT	Society of Worldwide Interbank Financial Telecommunications
TCP/IP	Transmission control protocol/Internet protocol
TLD	Top level domain
URL	Uniform user locator
VoIP	Voice over Internet protocol
WiFi	Wireless fidelity
WLAN	Wireless local area network
WWW	World Wide Web

INDEX

Abler, R.F.   16, 81
Abu-Lughod, J.L.   101, 103, 107
academic   206
  communities   87
  institutes   90
  interest   202
  knowledge   58
  networks   87, 152
  research   63, 78
access   20, 22, 24, 34, 35, 42, 60, 86, 114, 133,
      140, 145, 148, 151, 176, 177, 180, 182,
      185, 187, 193, 197, 201, 215
  websites   117
accessibility   33, 55, 56, 128, 141
activity, space   188
actors   42, 127
Adams, P.C.   7, 24, 39, 133–5, 137, 188
advantage, initial   103, 156
  initial and cumulative   8
  relative   162
  sustainable   68
advertising   94, 102, 114, 128
Africa   55, 73, 146, 171, 179
age, of information   9
Agnew, J.A.   33, 39
agriculture   57
  employment   55
  production   55
Ahnstrom, L.   77
airline
  network   145
  routes   144
  tickets   6
Akamai   153, 195
Alexa Research   133
Alexander, J.   24
Aliso Viejo   133
Alta Vista   116
Amazon.com.   117, 132–3
America   172, 174
  see also US
American   47, 72, 99, 100, 101, 103, 104, 106,
      107, 109, 117, 132, 156, 158, 164, 175,
      181, 187, 188, 191, 193
  cities   101, 131, 151, 155
  economy   156
  invention   156

leadership   156
lifestyle   174
market   159
metropolitan economies   75
mobile telephone   197
society   11, 155–6, 175
system   146
Americas   186
Amin, A.   44, 120
Amsterdam   146, 148, 159
Angel, D.P.   166
Anglo-Saxon countries   74
Anguilla   131
Anshelm, M.   173
Antigua   118
Antonelli, C.   58, 61, 67
AOL (America OnLine)   20, 98, 131, 133, 140
Aoyama, Y.   170, 172
APNIC (Asia-Pacific Network Information Center)
      186
Appadurai, A.   41, 47, 49, 121, 126
Arab–Israeli conflict   77
architects   126
architecture   76, 102
Argentina   122
ARIN (American Registry For Internet Numbers)
      186
Aristotle   6
Arndt, O.   61, 66
ARPANET   12, 88, 141, 207
art   94, 105, 177
Asia   47, 100, 146, 147, 148, 150, 159, 179, 186
  cities   148, 191
  countries   24
Asian Americans   200
Associated Press   187
AT&T   19
Atkinson, R.D.   34, 57, 73, 75
Atlanta   72, 144, 145, 149
Atlantic   74, 190
audio webcasts   200
Audretsch, D.   58–9, 61–2, 64, 73–5
Aurigi, A.   181
Austin   72, 76, 89, 106, 128, 189, 190, 193
Australia   46, 118, 165, 175
Austria   165, 167
automobiles (cars)   86, 127, 164, 174, 193

B2B   128–9
B2C   128–9
backbones   89, 144, 146, 148, 149, 151, 154,
      158, 159, 212
   capacity   147, 150, 154
   cities   145
   colonialism   155
   companies   141
   firms   144
   network   141, 145, 150
   providers   131
   system   143, 148
   traffic   141
   transit   140
Bahamas   122
Baltimore   193
bandwidth   5, 8, 35, 118, 131, 141, 143, 144,
      145, 147, 150, 153, 159, 197, 214
Bangalore   73–4
banking   14, 44, 93, 126, 127, 133, 163, 165,
      175, 186
   capital   103
Barcelona   99
Barlow, J.P.   2, 7, 25
Barr, B.   18
Barrie   32
barrier   25, 34, 40, 102, 120, 124
   factos   35
barrier-free   121
Barthes, R.   32
Barua, A.   127–30
Batty, M.   18, 36, 134–5
Beaverstock, J.V.   103
behavior   3, 15, 54
Beijing   100, 192
Belgium   165, 167, 175
Bell companies   20
   system   19
Bell, A.G.   77, 173
Bell, D.   3, 12, 13, 16
Ben-David, D.   59
Benedikt, M.   34
Beniger, J.R.   10
Benjamin, W.   134
Berkeley   88
Berlin   99, 146
Bermuda   131
Biegel, S.   87
biotechnology   71
   industry   58
Bird, J.   49
Birkinshaw, J.   62
BITNET   88–9, 151
Boisot, M.H.   2, 25–6
Bonn   100
book publishing   91, 110, 212, 215
books   2, 4, 6, 13, 24, 25, 29, 30, 36, 81, 86, 87,
      91, 104, 105, 108, 111, 129, 192, 211
bookstores   105
borders   194
   see also boundaries
Borgmann, A.   2

Borsook, P.   66, 67, 73, 74, 76, 89, 109, 110,
      132, 143, 145, 206, 212
boundaries   39, 40, 120, 210
   see also borders
Braczyk, H.J.   93
Braman, S.   2
Brazil   73, 122, 196
Brealey, R.   93
British   46, 104, 132, 133
   domain registration   96, 99
   Empire   175
   see also UK
broadband   96, 186, 198, 201, 213
   communications   192, 196
   ISP connection   215
   services   200
   transmission   159, 198, 199, 203, 205
broadcasts   18
   media   87
   services   174
Brooker-Gross, S.R.   56
Brunn, S.D.   34, 76–7, 128
Brussels   146
Bryson, J.R.   9, 57
bubble economy   206
Bureau of the Census   187, 192
business   17, 18, 50, 60, 77, 81, 90, 91, 103,
      104, 110, 113, 114, 123, 124, 130, 132,
      136, 140, 153, 156, 161, 163, 166, 170,
      185, 190, 193, 198, 200, 202, 206, 208,
      209, 212
   activity   120, 129;
   capital   103
   communications   170
   community   190
   correspondence   36
   driver   127
   environment   131
   hotels   126
   information   103
   knowledge   60
   management   186
   media   172
   meetings   138
   operations   91
   organizations   62
   types   198
   uses   187
Buttenfield, B.P.   135
Button, K.   133
Buy.com   133
buyers   129

Cable & Wireless   141
cable TV   5, 10, 13, 18, 20, 31, 44, 85, 104, 109,
      141, 157–8, 185, 202
   companies   20
   transmissions   198
caching   153
   servers   153
CAIDA   195
Cairncross, F.   27, 34, 215

California 67, 89, 104, 106, 108, 114, 143, 149, 172
Camagni, R. 59, 67, 68
Cambridge (UK) 73
Campbell, D. 50
Canada 118, 164, 165, 167, 175
capital 6, 12, 15, 24, 27, 31, 42, 46, 47, 50, 54, 56, 57, 58, 67, 73, 79, 102, 108, 111, 113, 119, 125, 126, 133, 162, 175, 192, 206, 209, 210, 211
  direction 76
  flow 67, 79, 120, 133, 210
  formation 65, 102
  industry 101, 113
  market 104, 207
  mature,75
  mobilization 60, 68, 111, 211
  region 100
  revolution 41
  sources 47
capitalism 11, 12, 50, 56, 209
capitalist 50, 126, 156
  national economies 76
  system 74–5
Carey, J.W. 55–6, 174
Caribbean 118, 119
casinos 118
Castells, M. 2–4, 10, 12, 15, 18, 20, 33–4, 39, 41–3, 49–50, 60, 64–5, 67–8, 95, 99, 102, 126, 129, 156, 162, 172, 176, 179, 181–2, 185–7, 190, 200, 206
Cayman Islands 122
CBD 102, 123, 124, 126, 131
  see also downtown
CcTDL 90
CDMA (Code Division Multiple Access) technology 196
cellular telephony 14, 77, 161, 166–7, 170–4, 195, 197
  industry 77
  radio 157
  subscribers 196
  see also mobile
CEMA 109
center 24, 47, 58, 65, 68, 71, 72, 73, 75, 95, 101, 105, 108, 123, 132, 133, 134, 157, 207, 210
  major 111, 185, 211
  of information technology 89
  of Internet 117
center-periphery 181
centrality 21, 90, 111, 123, 137, 145
centralization 210
CERN 88
Chandler, A.D. Jr. 11
channel 5
Chan-Olmsted, S.M. 21
chat 116, 138
Chicago 67, 71, 76, 101, 105, 108, 109, 132, 143–5, 148, 150, 189
China 74, 101, 154, 192, 195
  cities 192
  pattern 100

Christopherson, S. 107–8
Ciolek, T.M. 24
cities xii, 8, 22, 48–9, 54, 75–6, 85, 93–7, 100, 101, 102, 108–11, 114, 118–20, 123, 125, 126, 132, 141, 143, 145, 154, 161–2, 173, 176–7, 180, 181, 182, 188, 190, 191–2, 197, 201–2, 210–11, 212–16
city 22, 61, 75, 83, 95, 103, 104, 105, 109, 144, 146, 147, 150, 183, 189, 190, 191, 206, 214
  boundaries 75
  broadcasting 102
  level 195
civilization 174
Clark, G.L. 9
Clark, T.H.K. 127, 133
Cleveland 190, 191
Clinton–Gore administration 88
clusters 62, 64
code 37, 40, 61, 138, 186
  red worm (crv2) 154
codification 59
codified 1, 7, 26
  knowledge 9, 23, 26, 58, 123
Coe, N.M. 131, 175
Cold War 11, 12, 156, 174
collocation 140
  agreements 200
  center 207
  facility 149, 151, 152, 153, 214
  industry 149
Colorado 106
Colosource 151
commerce xii, 14, 33, 128, 133, 138, 165, 175, 177, 212
  companies 62, 90, 132
commercial 128, 132, 199, 208
  activities 127
  advertising 174
  domains 108
  firms 200
  information 115, 131, 197
  provision 128
  reach 56
  sites 114
  ties 78
  transactions 127
commercialization 23, 62, 206
Commercial Internet Exchange (CIX) 141, 148
commodity 2, 10, 14, 16, 41, 56, 58, 103, 121, 144
common carriers 19
communications 3, 20, 36, 40, 49, 87–8, 93, 113, 117, 132–3, 138, 151, 158, 161, 162, 172, 173, 176, 182, 186, 195, 202, 205
  appliances 16
  centrality 123
  culture 175–6, 196
  device 197
  hierarchy 213
  infrastructure 197
  inventions 202
  means 163, 195, 203, 213

communications (*continued*)
  media   56, 133, 165–7, 71, 172, 173, 176,
      182, 187, 195, 201, 203, 213
  modem   196
  modules   140
  networks   40
  products   213
  protocols   139, 196
  satellites   157
  services   176
  system   86–8, 102, 141, 180, 198
  technologies   55, 88, 124, 127, 176
  time   46
  tradition   175
communicative materials   1–4, 27, 81, 113, 208
community   137, 138, 162, 177, 190
  call centers   181
  memory   190
companies, computer and telephone   18
company   73, 76, 94, 104, 110, 123, 124, 129–2,
      187, 194, 206–7, 212
  acquisitions   206
  core-located   181
  headquarters   132
competition   75, 125, 162
competitiveness   59, 64, 68, 78
computer   2, 10, 12, 13, 15, 16, 18, 19, 20, 21,
      29, 36, 40, 50, 80, 85, 87, 88, 94, 103,
      109, 114, 117, 127, 134, 135, 136, 139,
      140, 153, 154, 156, 157, 161, 163, 171,
      186, 206
  communications   134, 137, 190
    networks   88
  companies   20, 129
  development   158
  hacking   208
  hosts   26
  images   135
  keyboard   209
  networks   85, 91, 136–7, 139, 177, 208
  ownership   181
  power   35
  products   129
  programs   136
  systems   132
  technologies   5
  terminals   127
  users   137
computerization   206
  global networks   137
  information   121, 134
  Internet   196
computing   104, 177, 206, 214
concentration   76, 77, 81, 85, 93–97, 99, 100,
      103–107, 109, 114, 118, 131, 150, 201,
      208, 211
  of producer services   79
Confucian society   199
conglomerates   208
connectedness   13, 14
connection   46, 59, 87, 128, 145, 148, 150, 196,
      197, 200, 292
  times   192

connectivity   14, 16, 132, 141, 177, 182
construction   126
consumer   132, 51, 130, 141, 149, 161, 209
consumption   xi, 8, 9, 10, 15, 22, 27, 30, 34,
      36–8, 41, 53–4, 56, 75–6, 79, 81, 85, 89,
      96, 98, 109, 111, 149, 163, 189, 190, 201,
      202, 214
  of information   185
  prices   21
  services   185
content   xi, 8, 17, 24, 34, 81, 86, 95, 98, 104,
      108, 111, 112, 138, 139, 141, 149, 177,
      197, 200, 211, 212
  business   91
  demand   113
  industry   199, 207
  production   94–5
contextuality   9, 180
continuous learning   60
control   22, 95, 103, 120, 162
  center   102, 104, 111, 211
  of information   214
  tasks   23
convergence   215
Cooke, P.   66, 67, 158
coplacement   49
copresence   49
copyright   9, 200
  laws   25
cores   162, 173
  areas   102
Corey, K.E.   65
Cornford, J.   44
corporate funds   62
corporations   128
CorpTech   94
Cortada, J.W.   11
Cortright, J.   75
Cosgrove, D.   32
country   41, 47, 48, 55, 68, 74, 75, 76, 81, 83,
      85, 90, 97, 100, 103, 118, 119, 120, 123,
      138, 146, 147, 153, 155, 156, 158, 159,
      161, 162, 163, 164, 165, 166, 167, 170,
      171, 172, 173, 174, 175, 176, 177, 179,
      194, 195, 199, 213, 214
  code   90
  government   63
  of innovation   68
Cowan, R.   7
Cox, K.R.   120
Crang, M.   30, 34, 54
creativity   177
Crevoisier, O.   66
criminal economy   182–3, 214
Cronin, F.J.   163
Cukier, K.N.   140, 155
cultural   10, 40, 47, 110, 118, 162, 173, 183,
      211, 212
  activity   12, 95, 104
  battles   15
  dimension   15
  domination   50
  expressions   4

heritage 76
identity 10, 41
imperialism 15
importation 44
industrial complex 102
messages 41
movements 15
national attachment 65
patterns 200
preferences 213
proximity 123
services 102
spaces 50
symbolism 15
value 15
culture 11, 14, 25, 27, 36, 76, 137, 172, 174,
    176–7, 185, 187, 209
  of information 15
cumulative advantage 68, 73, 104–5, 108, 111,
    112, 123, 210–1, 215
CUNY 88
Curry, J. 128
customers 93, 94, 124, 127, 128, 158, 194, 200,
    210
  locations 194
customization 129
cyber court 26
cyberspace xi, 30–1, 33–7, 41, 135–7
  -based world 39

Dallas 72, 76, 144, 145, 149
data 1, 2, 5, 6, 14, 15, 21, 86, 87, 113, 117, 123,
    140, 152, 153, 163, 167, 185, 188, 189,
    192, 208, 216
  bases 128, 131, 195, 208
  centers 152
  flows 154
  link 139, 140
  packet 195
  replication 153
  services 19
  sets 14
  traffic 83, 85, 87
  transmission 148
data-processing machines 18
David, P.A. 19
distance, death of 27
decentralization 162, 208, 210
dedicated networks 156
DENIC eG 100
Denmark 214
deregulation 156, 174
destination 48, 185, 187, 197
de-territorialization 44
Detroit 104, 191
developed 161, 180, 181, 195
  country 50, 54, 85, 120–1, 154, 162, 179–81,
      196, 199, 209
  economies 49, 57
  nations 163
  societies 12
  technologies 157

developing country 100, 121, 154, 179, 180,
    181, 195–6
development 19, 55, 65, 74, 75, 77, 88, 102,
    104, 123, 126, 157, 172, 174, 176, 186,
    190, 195, 196, 199, 205, 207, 210, 214
  countries 84
  of global capital 120
  policies 23
DeVol, R.C. 66–8, 72–3, 75, 107, 189
diffusion 89, 111, 155, 156, 157, 167, 171, 172,
    191, 196, 200, 213
  of computers 158
  of Internet domains 97
digital
  bit formats 5
  communication 83
  divide 23, 43, 73, 85, 161, 176
  envoy 195
  gaps 181
  information storage 82
  information 41
  phone service 18
  switching 157
  technology 18
  transmission technologies 197
  world 36
digitization 108
  of information 25
direct international dialing 163
direct marketing 194
disks 111, 211
distance 12, 32, 34, 49, 57, 62, 135, 137, 174,
    181, 195, 215
distribution 12, 20, 22, 27, 82, 91, 95, 150, 151,
    153
  centers 138
  of information 24
distributor (website hosting computer) 92
  wholesale/retail 91
division of labour 80
Dodd, N. 119
Dodge, M. 24, 30, 34–6, 85–6, 88, 124, 133,
    186
domain 94, 95, 96, 97, 98, 105, 109, 114, 118,
    188
  name 26, 90, 91, 94, 98, 99, 100, 101, 110,
      179, 188, 195, 212
  registration 89, 91, 94, 95, 97, 99, 100, 145,
      199
  system (DNS) 89
domestic 26, 27, 67, 78, 123, 125, 172, 174,
    182, 183, 213
  accumulation 78
  business 123
  and foreign banks 123
  Internet economy 96
  investment 79
  life 48
  market 173, 175
  offices 126
  -regional 59
  spatial language 137
domination 33

Doron, A. 113
downstream ISPs 140
downtown 126, 131–2
  see also CBD
Driver, S. 104
Duncan, J. 32, 40
Duncan, N. 32
Dunning, J.H. 9, 57–8, 62
Düsseldorf 146

Eade, J. 43
early bird, advantage 199
EARN 88, 152
European, east 74, 151, 171
ebay 133
e-commerce 22, 36, 113, 127, 136, 176, 186,
    192, 197, 206
  customers 194
economic 9, 10, 14, 26, 31, 36, 110, 173, 176,
    177, 183, 206, 208, 209
  activity 8, 86, 94–5, 126, 127, 138, 162, 186,
    198, 209, 214
  actors 51
  areas 143
  cores 101
  definition 127
  development 9, 74, 76, 127, 162, 192
  dynamism 75
  growth 175
  knowledge 9
  nature 47
  power 42, 105, 165
  process 5
  resource 9
  specialization 92, 93, 112, 117, 138, 162, 212
  stability 71
  structure 176
  vitality 190
Economist, The 19, 20, 21, 138, 141, 153, 194–8
economy 1, 15, 57, 104, 127, 170, 182, 183,
    213, 214, 215
  idea-based 57
  informal 182
  of scale 93
education 7, 9, 23, 57, 74, 114, 116, 171,
    180–1, 192–3, 199
educational materials 114
Edwards, T.M. 214
Ein-Dor, P. 179
Elazar, D. 175
electricity 19, 21, 86, 206
electronic,
  agora 137
  bits 18, 53, 120
  communications 61, 117
    medium 182
    system 83
  connections 124, 127;
  consumption 205
  data interchange 127
  data processing 158
  exchanges 42
  formats 1

forms 31
frontier 26
global stock trading 206
information 5, 6, 14, 29, 31, 35, 40, 88, 93,
    133
  media 163
  packets 207
  source 83, 185
  system 207
inter-computer 87
learning 202
library 86
mass-media 164
media 46, 84, 164, 165, 208
  outlet 215
  processing 8
  products 111, 211
  signals 5, 198
  space 29, 36, 44
  strands 128
  study xii
  submission 215
  telephone switches 87
  transfer 127
  transmission 2, 102
  -virtual information 22
  volume 133
  world 215
Elkin-Koren, N. 25
e-mail 5, 9, 24, 83, 86–7, 89, 94, 113, 127, 141,
    181, 186–7, 190, 198, 208, 216
end-user 20, 198, 215
energy 19, 127, 149, 153, 208
England 41, 90, 97, 99, 131, 136, 159, 171, 173
Engstrom, J. 106
enterprise 66, 80, 212
entertainment 4, 17, 48, 106, 107, 108, 115–6,
    129, 163, 172, 190, 199, 208, 214
entrepreneurial 71, 110, 211
entrepreneurship 13, 54, 58, 60, 65, 72, 73, 75,
    76, 107, 108, 111, 190, 207, 211
Entrikin, J.N. 32, 40
environment 34, 67, 122, 134, 190, 205
e-purchase 36, 192
  companies 73
  see also e-commerce
Ericsson 77, 173
ethnicity 181
ethnoscapes 41
Europe 73, 74, 88, 98, 132, 146, 148, 150–1,
    155, 157, 159, 170, 179, 186, 196–8, 203,
    213
European,
  carriers 197
  centers 73
  cities 148
  Commission 10, 13
  companies 89
  countries 165, 171, 176
  firms 61
  international traffic 155
  lines 147
  market 176

nations   90, 158, 167, 175, 176
  settlement   11
  Union (EU)   55, 121, 175, 176
Evans, P.B.   16
exchange   54, 56, 59, 60, 61, 77, 93, 95, 102,
    109, 127, 139, 208, 211, 212
  of information   68
  of knowledge   68
Exodus   140, 153, 205
expansion   89, 93, 119, 123
experience   2, 31, 33, 56, 57, 79, 134
expertise   71, 91, 95, 103, 110, 123
Explorer   88
export   173, 175
expressways   144
extensibility   137, 188
Eyescream   116

Fabrikant, S.I.   135
facilities   198, 206
fashion   48, 102
fax   6, 15, 18, 86, 161, 163, 166–7, 170–2, 185
  services   14, 19, 157, 166, 182, 213
  transmission   5, 83, 139
Feather, J.   14
federal capital   214
  government   98
  regulation   197
Federal Communications Commission (FCC)
    20–1, 198
Federal Express   85, 105, 189
Feldman, M.P.   3, 29, 58–62, 64–5, 68, 73–5
Felsenstein, D.   64
Ferguson, J.   40, 49
fiber optics   18, 131, 141, 143, 157–8, 200
film   8, 134
  companies   18
  industry   107
  production   82, 104
  see also motion pictures; movies
finance   xii, 16, 102, 103, 104, 104, 108, 111,
    114, 117, 117, 133, 138, 211, 212, 215,
    67, 93
  industry   8, 92, 93
financial,
  activities   207
  business   124
  capital   120
  centers   73, 77, 102, 124
  centrality   103, 111
  companies   207
  conditions   206
  cores   121
  crisis   206
  engineering   102
  information   103, 114, 193
  institutions   121, 124
  products   104
  sector   85, 99
  services   104, 208
  settlements   148
  tacit knowledge   123
  technology   123

tools   60
financing   62, 126
finanscapes (finances)   41
Finland   65, 77, 165, 167, 170, 173, 214
Finney, M.   104, 107
Finnish   76
FIRE   92, 102, 103, 104, 107, 111, 211
firms   58, 59, 61, 62, 71, 77, 79, 94, 110, 123,
    197, 210, 212
  policy   50
first-mover advantage   67, 111, 131, 210
Florida, R.   60–1, 66, 68
flow   8, 9, 13, 22, 24, 30, 37, 41–3, 47, 50,
    54–6, 67, 120, 122, 123, 133, 144, 153
  global capital   120
  of information   41, 138, 209
  of technology   42
  patterns   89
fluidity   34, 86
foreign,
  investment   79
  direct investment   121–2
  countries   180
fragmentation   50
France   85, 121, 132, 165, 167, 173, 194, 208
Frankfurt   123, 146
French   127, 194
  domains   99
  government   74
Frenkel, A.   68
friction of distance   55
Friedmann, J.   41
FTP   87
furtive money   122
fusion   15, 44, 47, 48, 49, 102, 107, 108, 124,
    133, 136, 138, 183, 187, 208, 210
  processes   18, 21, 56
FX (foreign exchange)   123, 126

gambling   26, 129, 194
Garnsey, E.   67
Garreau, J.   124
gateways   77, 79, 150
Gelernter, D.   136
Geller, P.E.   25
gender   181
general purpose, information machines   198
  technologies (GPT)   21, 86
generic TLDs (gTDLs)   90
geographer   15, 22
geographic   3, 212
  bias   94
  communities   137
  contours   143
  interface   136
  language   133
  metaphors   138
  movements   175
geographical   27, 33, 39, 53, 95, 110, 111, 136,
    156, 161, 165, 167, 171, 172, 173, 175,
    176, 215
  accuracy   194
  address   197

geographical (*continued*)
  aspects   30, 113
  attributes   131
  behavior   54
  breakdowns   121
  change   30
  concentration   121, 177, 180, 181
  concept   81
  conception   54
  convergence   111, 210
  cores   207
  decentralization   68
  definition   75
  destinations   47
  difference   132, 216
  diffusion   88
  dimensions   30
  distribution of knowledge   73
  distributions   73
  entities   209
  expansion   126
  extent   31
  fineness   94
  fluidity   26
  foci   83
  framework   59
  groups   165
  identification   197
  images   135, 136
  information system (GIS)   30
  landscape   122, 126
  language   31, 37, 135, 137
  limits   59
  location   79, 176, 193, 194, 194, 197, 210
  notions   136
  ordering   198
  organization   108
  origins   47, 121
  path   212
  pattern   8, 81, 105, 111, 114, 154
  perspective   86, 89, 113, 195
  proximity   62, 110, 175, 212
  ramification   113
  range   140
  requirement   122
  separation   107
  setting   60
  sources   24, 157
  space   25
  spread   94, 165
  study   xi, xii, 22
  terms   136
  territory   214
  travel   158
  trends   121
geographically,
  adjacent   122
  concentrated   23
  distributed   86
  more flexible   67
  varied flows   50
geographies 27, 30, 31, 53, 139, 140
  of information technologies   80

  of infrastructure   21
  of production   9
geography   1, 9, 21, 30, 31, 36, 40, 43, 61, 95,
      113, 127, 131, 138, 215
  of information   xi, xii, 1, 29, 37, 42, 80, 81,
      205, 208, 215
  of innovation   29, 54, 74, 79
  of knowledge   133
  of networks   30
  of telecommunications   xii
  restricting forces   25
  real   36
geolocation   26, 193, 194, 197
Germany   90, 96, 98–9, 121, 132, 133, 158–9,
      165, 170, 173, 196
  cities   146
  leadership   146
  pattern   101
Gibbs, D.   23
Gibraltar   122
Gibson, W.   136
Giddens, A.   32, 40, 42, 44, 46–7, 49, 56
Giles, C.L.   83, 115
Gillespie, A.   34, 74, 104, 162, 181
global   9, 24, 25, 27, 29, 31, 33, 43, 46, 48–50,
      55, 56, 102, 103, 104, 121, 123, 124, 125,
      128, 132, 134, 138, 147, 156, 157, 159,
      162, 177, 182, 183, 188, 209, 210, 213,
      214
  activity   67
  affairs   43
  assets   121
  audience   47
  business   175
  capital   104, 120, 123, 126
  capital centers   101, 120
  capital flows   101, 120, 121, 122
  capital investment   120
  capitalism   50
  center   27, 42, 102, 123
  centrality   101
  cities   41, 42, 46, 84, 101, 125, 131, 154
  concentration   112
  contexts   125
  control   43
  cores   55, 101, 121, 172
  cultural symbols   102
  culture   106, 183
  cyberspace   137
  decentralization   97
  dispersion   101
  distribution   50, 68, 79, 208, 209, 210
  economic cores   77, 179
  economic   173
  economy   55, 65, 93, 104, 105, 120, 175, 199
  electric flow   121
  estimate   83
  finances   93
  financial activity   210
  financial assets   121
  financial business   102
  financial center   93, 94, 99, 123
  financial economy   125, 207

financial markets   93
financial status   99
financial system   99
financial–technology crisis   207
flows of capital   42
flows   27, 41, 42, 42, 44, 47, 121, 125, 126, 209
formations   43
imprints   47
information   213
information networks   48, 134
information production   182
information space   103, 211
information system   25
innovative   174
interaction   46
international traffic   85
investments   46
knowledge   57
leader   97, 101
leadership   99, 102, 105, 163, 165, 166, 167, 173, 192, 213
localism   49
location   111, 210
maritime trade gateway   175
market areas   128
markets   117
media   44
mobility   73
modes of domination   50
movement   47, 48
nature   132
network of computers   85
networks   79, 137, 138, 197, 210
orientation   78
per capita production   82
perspective   96
power   50
production   83, 118
R&D, activities   60
  center   65
reorganization   156
scale   24, 41, 67, 119, 120, 120, 124, 159, 186, 203, 213
sense of place   40
service   123, 124
social networks   49, 138
sources   50, 209
space   15, 44, 122, 138
spatial, barriers   124
  language   137
specializations   125
system   73
telecommunication   56
telecommunications hub   175
transfer of information   215
transmission   134
urban hierarchy   125
venture capital   24, 78
view   96, 146
virtual   15
Global Reach   171
globality   182

globalization   13, 16, 39–44, 47, 57, 62, 65, 67, 75, 79, 92, 93, 101, 121, 124, 128, 155, 162, 167, 176, 186, 210
  of innovation activities   59
global-local   50
global-virtual   41, 49
globally   31, 34, 121, 159, 175, 207, 213, 214
  accessible   86
  leading city   119
  relevant   182
  spread   59
glocalisation   44
Goddard, J.   1, 16, 47
Goldstein, H.   58
Goodchild, M.E.   91
goods   54, 58, 163, 193
gopher   87
Gordon, D.   120
Gorman, S.P.   36, 86, 131, 139–40, 145, 148, 152–3
Gottlieb, P.D.   57, 73, 75
Gottmann, J.   11, 15
government   23, 24, 26, 60, 64, 77, 101, 138, 144, 193, 194, 199, 200
  agencies   89
  control   44
  programs   79
governmental   19, 88, 90, 131, 181, 194
  intervention   175
  involvement   78
  organization (NGOs)   128
  policies   20, 120
  programs   79
  structure   176
GPS (Global Positioning System)   197
Graham, S.   15, 21, 22, 29, 31, 35, 36, 41, 74, 85, 93, 95, 97, 99, 126, 131–2, 137, 139, 162, 181, 207
graphical transmission   6
graphics   5, 14, 21, 86, 118
Gregory, D.   31, 44–5, 126
Grimes, S.   181
grocery   129
Grote, M.H.   123–4
growth   179, 191, 198
Grubesic, T.H.   140–1, 144–5, 148, 152, 199
GSM technology   196
Gulia, M.   137
Gupta, A.   40, 49

Hackler, D.   68
Haifa   79
Haiti   179
Halal, W.E.   11, 15
Halavais, A.   24
Halbwachs, M.   49, 137
Hall, P.   64–5, 95
Hamburg   146
handling, of information   79
Harasim, L.M.   134, 136–8
harbours   175
hard drives   82

hardware   20, 53, 88, 103, 105, 106, 109, 111,
     129, 132, 140, 191, 211
   companies   20
Hargittai, E.   171
Harris, R.   86
Harvey, D.   32, 33, 37–8, 39, 44, 46, 56–7,
     124–6
Haug, P.   106
headquarters   76, 104, 126, 133, 138, 175
health   116
Helleiner, E.   120–1
Helpman, E.   21
Helsinki   77
Hepworth, M.E.   13, 16, 21, 29
Hewson, M.   43
Heydebrand, W.   105
hierarchical   89
   organizations   187
hierarchy   182, 190
high innovative content   76
high-learning   75
high-risk capital   67
high-speed transmission   198
high technology   75, 93, 95, 97, 98, 106, 108,
     114, 117, 145, 206, 208, 212
   business   67
   centers   71, 73, 77, 79, 144, 190, 210
   cities   110, 212
   companies   94
   concentration   76
   economy   75
   entrepreneurship   78
   hub   170
   image   127
   industrial parks   23
   industry   9, 13, 24, 54, 64, 67, 65, 79, 93, 94,
     109, 110, 144, 145, 190, 211
   information technology   110, 211
   innovation   74, 76
   innovative firms   62
   output   75
   parks   126
   production   75, 189
   R&D   23, 24, 74, 78
   regions   68, 110
   sector   77, 79
   training   78
   workers   71
Hill, R.   127
Hillis, K.   xi
Hillner, J.   68–72
historical, developments   156, 174
   industrialization   61
Hodge, D.   55
Hodgson, G.   9
Holloway, S.I.   25, 36
Hollywood   99, 102, 106, 107, 108, 112, 118,
     211
home   19, 198
   countries   65
   users   185
   video   192
Homebrew computer club   190

homogenization   50
Hong-Kong   24, 74, 122, 123, 126, 165–7, 170,
     172, 175, 181, 196, 199, 208
Hopkins, J.   134–5
Horrigan, J.H.   190
host   90, 179
   registration   91
Host, S.   173
hosting   186
   computers   86, 95
   facilities   152, 158
   server   89, 118
Hotz-Hart, B.   57, 59
households   15, 18
Houston   76, 181
Howells, J.   7
HTML   88
'hub and spoke'   123
hubs   30, 31, 42, 89, 148
Hugill, P.J.   xi
Huh, W.K.   100, 154
human,
   activity   64
   agents   50
   beings   50, 82
   dependency   55l
   identity   40
   interest   43, 86
   products   8
   relations   137
   resources   65, 68
   rights   48
hyperspace   49

IANA   186
IBM   20, 130, 131, 141
icons   4, 15
ICP companies   188
ICT   30
ideas   2, 3, 25, 57, 58, 60, 79, 134
   new   60, 65
identities   40–1, 44, 49–50
ideoscapes (ideologies)   41
Ilchman, W.F.   32
Illinois (Chicago)   89
images of places   49
ImagiNation network   136
imagination   33, 34, 56
imaginative   34
imagined space   33, 34, 36, 55
immigration   65
immovable places   135
IMODE mobile Internet   197
import, of information   138
identity   26
India   74, 179, 196
individual   40, 91, 110, 185
individualism   172, 188
industrial, area   76, 131
   cores   210
   development   77
   experience   73
   organization   61

parks 64, 132
plant 47
policies 175
production 24
revolution 55, 158
society 11
industrialization 48, 100, 176
industry 21, 62, 65, 67, 74, 76, 111, 157
  cities 191
  of electronic information production 54
  plants 92, 133
  production 68, 125
in-flows 84
information xi, xii, 1–9, 25, 30, 34, 41, 42, 44,
    46, 47, 50, 54, 55, 56, 57, 59, 78, 80, 81,
    95, 102, 107, 110, 111, 113, 116, 117,
    119, 123, 127, 128, 129, 133, 134, 141,
    144, 154, 155, 159, 161, 163, 176, 177,
    180, 185, 188, 191, 194, 195, 198, 202,
    206–210, 212–5
access 176
activities 16, 105, 128
age 15, 29, 30
and communications system 87
and communications technology (ICT) xi, 16,
    26
and knowledge 1, 9
  globally transmitted 41
  protection 2
and technology 53
bodies 113
business 16, 20, 33, 54, 102
capital 103
channel 21, 182, 213
city 109, 112, 131, 132
classes 54
classification 4
companies 20
consumers 183
consumption 22, 103, 105, 111, 161, 163,
    177, 214, 207
designated 5–6
devices 14
distribution 22, 94
dominated society 14
economic geography 21
economy 1, 75, 95, 108, 110, 111, 112, 144,
    145, 146, 155, 156, 174, 190, 211, 212,
    215
employment 13
exchange 15, 128, 47, 86
facilities 155
files 113
flows 25, 84, 134, 153, 192
forms 5, 21, 86, 102
gateway 84
generation 10
geography 29, 31
handling 53, 93, 210
identification 88
industries 97, 98 102, 108–11, 118, 156, 174,
    206, 205, 211, 212
items 87, 193, 208

leadership 103
library 186
low 25
machines 134
materials 86
media 213
networks 104, 119
operators 20
outlets 19
perspective 127
politics 22
processing 10
producer 24, 89, 91, 92, 205, 211
production 5, 11, 12, 13, 14, 19, 22–3, 26,
    27, 30, 42, 53, 54, 74, 79, 80, 81, 117,
    132, 133, 139, 145, 146, 150, 153, 155,
    156, 158, 159, 177, 185, 188, 180, 190,
    191, 192, 208, 210, 211, 214, 216
products 26, 129
pure 5–6
receivers 5
revolution 18, 27, 41, 158, 208
searches 187
service-packages 20
services 15, 20, 23, 37, 176
society xi, 1, 9–15, 170, 78, 84, 214
space 102, 111
specialists 86
storage 5, 82, 86
superhighway 136
system xi, xii, 5, 21, 36, 50, 83, 86, 87, 88,
    113, 129, 138, 148, 186, 193, 206, 207,
    209, 208, 212
technologies 12, 13, 53–80, 87, 120, 124,
    131, 139
technology 5, 7, 8, 12, 16, 27, 29, 30, 36, 41,
    53, 85, 97, 102, 109, 111, 155, 182, 188,
    208, 209, 210
  center 68
  industry 212
  innovation 75
  regions 68
  workers 127
tools 128
transfer 3
transmission 6, 12, 96, 133, 135, 136, 195
types 17, 18, 105, 108, 109, 127
undesignated 5
use 5, 192
values 15
volumes and origins 81
work 12, 13
informational, economy 43, 65, 95
  products 91
information-based, phenomena 93
  services 182, 213
  society 13
information-intensive industries 114
information-poor 181
information producing 212
  places 138
information-rich 181
  society 12

INDEX

information services   185
  sharing   46
  system   198
informative interactions   127
informatization   10
infoscapes   41
infosplit   195
infrastructure   16, 17, 18, 19, 21, 22, 35, 36, 77,
    79, 93, 94, 130, 141, 144, 149, 154, 162,
    164, 166, 174, 176, 181, 182, 192, 207,
    207, 210, 214, 215
  business   131
  production   22
in-house tacit knowledge   58–9
initiatives   79
innovating firm   58
innovation   1, 2–4, 8, 10, 22, 24, 53–7, 62–5,
    67, 68, 74, 75, 77, 79, 80, 94, 104, 109,
    110, 155–8, 162, 175, 190, 191, 202, 203,
    205, 206, 209, 211–3, 216
  chain   67
  cycles   80
  diffusion   203
  networks   59
  process   7, 58, 59, 61, 64, 68, 207
  products   207
  rounds   68
innovative   177, 205, 209
  firms   58–60, 62, 68
  industrial R&D   63
  industries   80
  information   119
  information technologies   210
  knowledge   3, 23, 25, 27
  production   67
  products   3
  research   23, 64
  technical labour   67
insurance   16
integrated   19, 183, 198, 209, 213
integration   4, 18, 36, 44, 48, 67, 80, 156, 162,
    187, 190, 206
Intel   20
intellectual amplifiers   15
  creativity   10
  piracy   26
  property rights   25
  property   27
  rights   194
intelligence   128
intelligent corridors   65
interaction   14, 26, 34, 36, 37, 41, 44, 49, 74, 80,
    85, 129, 134, 135, 140, 163, 188
interactive,
  exchange   185
  information system   86
  television   200
interactivity   128, 129
inter-computer communication system   87
interconnected   159, 186
interconnection   148
intercontinental   146, 179, 214;
  backbones   159

communications   146
connections   146
digital divide   146
inter-firm relations   74
inter-metropolitan,
  backbones   146
  bandwidth   214
International Telecommunication Union (ITU)
    163, 166
international   26, 43, 146, 159, 175, 177, 180
  and interregional borders   57
  backbones capacity   146
  backbones   148, 155
  bandwidth   146
  banking center   175
  banking   125
  banks   120
  boundaries   13
  calling   165, 167, 175
  calls   174
  commerce   146
  communications   77
  country code   90
  digital divide   179, 180
  direct dialing   161, 165
  exchange   55
  finance center   93
  financial markets   120
  flows   119
  gateway   103
  interaction   47
  interbank   121
  Internet capacity   155
  ISPs   140
  junction countries   172, 176
  junction   165, 167, 175
  level   65, 95
  lines   147
  monetary fund   121
  movements   55, 121
  nature   126
  organization   163, 165
  phone calls   10
  space   54
  telecommunications   55, 120, 125
  telephone calling   171
  telephone calls   170
  telephone system   19
  telephony   13
  tourism   55
  trade   59
  traffic   55, 156
internationalization   23, 121
internationally   176, 186
Internet   xi, xii, 1, 2, 5, 6, 8, 10, 12, 13, 17–22,
    25 26, 30, 31, 36, 41, 51, 59, 63, 65, 80,
    83, 84, 85–9, 103, 124, 127, 128, 133,
    135, 157, 161, 163, 166, 167, 170–3, 176,
    177, 179, 185, 197, 198, 199, 201, 203,
    205–8, 210, 211, 213–5
  backbones   141
  business   132
  centers   181

communication 198
content 138
connection 197
consumption 182, 213
connected 198
culture 60
data 197
development 205
economy 127, 129, 132, 152, 189, 205, 214
expertise 98
hotels 149
industry 64, 67, 93, 104, 149, 179, 199, 206, 210
information 48, 198
  industry 211
  management 153
  production industry 205
infrastructure 130, 150, 181, 213
Internet2 152
language 138
markets 191
network 154, 186, 141
penetration 100, 105, 170
production,192, 215
  centers 114
  industry 97
products 141
projects 206
providers 19
registries (LIRs) 186
service 20, 192
service providers (ISP) 20, 131, 140
site 24, 59, 93
subscribers 202
system 19, 20, 21, 97, 193, 207
technology 14, 124, 159, 191
telephone technology 20
telephones (iMODE) 196
traffic 141, 148
transmission system 139, 158
transmission 91, 195, 197
travel services 37
use 14, 47, 113, 138, 194, 201, 208
website 182, 211
Internet.com 189, 191
intra-city flows 123
invention 65, 67, 77, 157, 174, 202
investment 63, 66, 67, 67, 78, 126, 206, 207
  market 125
  space 46
IP address 89, 129, 186, 193, 195
Iran 138
Iraq 179
Ireland 65, 170
Ireland, J. 93
ISBN 91
ISC (Internet Software Consortium) 178–9
ISDN 198
island nations 179
ISP 148–9, 151, 153, 158, 191, 197, 198, 215
Israel 65, 72, 74, 77, 91, 179, 208
  economic system 78
  government 78

high-tech 77–8
IT 61, 62, 64, 75, 139, 172
  R&D 74, 109
Italy 41, 132, 157, 165, 173
IXs 151

Jaffe, A.B. 62
Janelle, D.G. 46, 55, 56
Japan 24, 74, 121, 122, 140, 158, 165, 167, 170, 172, 179, 203, 206, 213
Japanese 41, 46, 74, 84, 197
Jerusalem 79
Jewish 107
  communities 77
Jin, D.J. 61
job creation 57
John Goddard 29
Johnson, B. 6, 7, 9, 25, 180
Jordan 179
journals 13, 86, 87
J-phone 197
junction countries 166, 170

Katz, P.L. 13
Kellerman, A. 5, 9, 12, 14–6, 18, 21, 23, 29, 31–3, 43, 54–6, 67, 77–8, 85, 87–9, 93, 101, 105, 109, 114, 120–1, 124–5, 139, 141, 154–8, 161–2, 170, 173–5, 179, 187, 188
Kenney, M. 128
Keohane, R.O. 159
Kim, H. 100, 154
Kim, Y.H. 41, 95, 104
King, A. 126
Kinky Komputer 190
Kipnis, B.A. 79
Kirsch, S. 33, 57
Kitchin, R. 14, 15, 24, 30, 34–6, 85–6, 88, 124, 162
Korea Network Information Center (KNIC) 188, 191, 202
knowledge 2–9, 12, 13, 22, 23, 25, 26, 27, 37, 38, 50, 53, 54, 57, 64, 65, 66, 68, 74–9, 79, 87, 91, 95, 103, 109, 111, 131, 132, 141, 177, 180, 186, 209–11, 216
classification 6
creation 110
development and spillover 63
economy 170
exchange 3
externalities 59
gap 176, 177
industry 97
new 65
producers 51
producing cities 131
production 8, 215
resources 58
society 177
sources 58
spillover 59, 62, 64, 79, 210
  in-region 59, 61

INDEX

knowledge (*continued*)
  transfer 62, 78
  types 58
knowledge/innovation formation 60
Knox, P.L. 41, 48
Kobrin, S. 120
Kopomaa, T. 195
Korea 96, 97, 138, 154, 172, 172, 188, 192,
    196, 200, 202, 203, 208, 213
Korean 154, 185, 191
  distribution 100
  society 199
Kornbluh, K. 200
Kotkin, J. 94, 102, 104, 108, 131
Kwan, M.P. 49, 135, 188, 192

labour force 16
labour 12, 47
Lakshmanan, T.R. 84
landscape 1, 32, 34, 68, 135, 136
  simulated 35, 136
Langdale, J. 123–4
languages 9, 25, 41, 99, 102, 110, 113, 134–6,
    170, 181, 194, 212
laptop computers 186, 196
Lash, S. 54, 134
Latin America 146, 148, 150, 179
Latour, B. 7
Laulajainen, R. 93, 121, 123, 126
Lawrence, S. 83, 115
LBS 198
leader 76, 95, 156, 162, 172, 174, 190
leadership 73, 75, 93, 96, 97, 100, 102, 103,
    103, 104, 105, 107, 108, 110, 114, 145,
    148, 150, 152, 154, 155, 156, 159, 164,
    165, 166, 167, 170, 171, 172, 173, 174,
    175, 182, 189, 190, 199, 201, 203, 213,
    214
leading 81, 105, 154, 156, 171, 172, 173, 175,
    176, 190, 199, 201, 203, 213
  areas 149, 161, 189
  centers 99, 100, 111, 210, 211
  cities 98–9, 101, 109, 112, 113, 133, 148,
    152, 188, 189, 191, 192, 210, 211
  core 155
  country 99, 100, 151, 170, 175, 199, 202
  economy activity 117
  example 98
  high-tech center 99
  information, cities 211
    industry 110
  Internet industry 99
  nation 161, 164, 165, 167, 170
  region 175
  role 155
  specializations 117
  states 106
  US cities 149
learning 2, 7, 60, 61, 68, 74, 163
  region 61
leased lines 85
Lefebvre, H. 32–3, 37, 56, 122
legal 194, 208

procedures 9
services 60
systems 26
legislation 200
Leinbach, T.R. 13, 34, 76–7, 124, 128
leisure 117, 172
  information 191
Lemos, A. 137
Lesotho 170
Lessig, L. 200
letters 2, 7
Li, F. 21–2, 29, 34–6, 41
liberalization 15, 23
Liberia 179
library 87, 216
license 25
licensing 64
Liechtenstein 122
listservs 190
literacy 164, 171
Litman, B. 21
local 29, 39, 43, 44, 46, 47, 65, 121, 125, 186
  activities 40
  call 20
  cultures 50
  economy 22
  information 193
  ISP 148
  media 25
  networks 197
  space 42
  specializations 22
  telephone market 20
  views 177
local–global dialectic 44, 46
localities 42, 44, 47–8, 125, 126
localization 124
location 7, 21, 35, 40, 42, 43, 54, 55, 68, 77, 86,
    92, 97, 111, 113, 122, 123, 124, 125, 127,
    132, 133, 137, 140, 144, 145, 146, 148,
    152, 153, 162, 172, 193, 194, 197, 198,
    210, 215, 216
  aspects 83
  of domains 191
  quotients 75
locational 9, 26, 43, 46, 65–6, 156, 175
  advantages 125
  consideration 153
  factor 65
  flexibility 171
  patterns 37
  perspective 92, 208
  requirements 149
  resource 55
  utilities 56
location-based 194
  services (LBS) 186, 197
location-dependent 138
location-free 138, 194, 212
locations 37, 47, 98, 108, 119, 120, 122, 123,
    124, 131, 133, 135, 138, 140, 156, 158,
    159, 194, 195, 212
Locksley, G. 91

Loewy, M.B.   59
London   42, 46, 85, 93, 96, 97, 99, 101, 104,
        109, 114, 120, 122, 123, 125, 126, 131,
        146–8
Long Beach   107
long-distance calling   174
Los Angeles   65, 75, 88, 89, 96, 97, 99, 101, 109,
        111, 112, 118, 126, 132, 143, 145, 149,
        150, 152, 153, 159, 189, 190, 201, 211,
        214
Lou, B.K.   10, 138, 172, 192, 205
Luger, M.   58
Luke, T.W.   43–4
Lundvall, B.A.   6, 7, 9, 25, 65, 74, 109, 180, 211
Luxembourg   126
Lycos   115, 116
Lyman, P.   81–4, 136, 192
Lyon, D.   12

Ma'ariv   114–5
machines   19
Machlup, F.   2, 7, 13
macromedia   131
Madrid   99
magazines   2, 41, 91, 103, 104, 107, 108, 111,
        192, 211
  publishers   24
Magic Cap   136
magnetic media   84
  storage   82
Maillat, D.   66
Makridakis, S.   19, 20
Malaysia   65
Malecki, E.J.   2, 21, 36, 53, 57, 59, 61, 68, 76, 86,
        129, 139–40, 143–6, 148–50, 152–3
Malmberg, H.   61, 64
management   12, 122, 124, 125, 129, 187
  of global capital   125
managerial elites   42
Manhattan   85, 105, 108, 112, 207, 211
manIpulation   27
manpower   47, 93, 94, 124
Mansfield, E.J.   62
manufacturing   57, 108, 111, 211
maps   134, 135, 136
Marable, L.   201
marginalization   182
market   22, 63, 64, 68, 71, 77, 78, 109, 167, 211
  activity   128
  development   67
  domination   74
  forces   23, 120
  growth   119
  knowledge   65, 66
  motivation   64
  place   103
  research   60
  success   57
  value   182
marketing   60, 64, 66, 94, 102, 127, 155
  system   174
Markoff, J.   207
Markusen, A.   61, 64, 75–6

Marshallian industrial district   61
Martin, W.E.   206
Martin, W.J.   11, 13, 15
Marvin, C.   56
Marvin, S.   21, 29, 56, 74, 85, 93, 95, 97, 99,
        126, 131–2, 139, 162, 181
Maryland   106
Maskell, P.   61, 64–5, 74, 76–7, 109, 211
mass,
  consumption   174
  media   49, 95, 133, 163
  production   64
Massachusetts   67, 89, 106
Massey, D.   32–3, 40, 46, 48–9
Mastercard   121
Masuda, Y.   10, 12
material   1, 38, 87, 209
  consumption   10
  entity and product   32
  information products   211
  objects   7
  or virtual   50
  organization   42
  place   134
  power   10
  productive force   38
  products   6, 14, 16, 25, 36, 57, 111, 182, 213
  space   29, 36
  spatial practice   33, 37
  support   188
Mayer, H.   75
MCI Worldcom   131, 141
McIntee, A.   149–50, 152
m-commerce (mobile commerce)   197
means, of telecommunications   163
measure innovative output   75
Media Matrix   114, 117
media   xi, 2, 7, 14, 15, 17, 18, 21, 22, 30, 50, 93,
        94, 95, 102, 104, 105, 108, 111, 134, 135,
        137, 161, 186, 208, 209, 210, 213
  industry   98–9, 102
  non-electronic   161
mediascapes   41
mediating   55
  forces   38
mediation processes   134
mediators   50
Mediterranean center   74
Megalopolis   15, 143
merchants   194
mergers   20
Merrifield, A.   32–3, 43, 122
messages   4
metaphor   136
metropolis   147
metropolitan   42, 49, 59, 80, 144, 146
  area   61, 68, 77, 97, 101, 109, 132, 152, 181,
        189, 206, 211
  cities   145, 202
  decentralization   102
  exchanges   148
  level   147, 181, 194

metropolitan (*continued*)
  region   59, 131
  spatial organization   123
Mexico   122, 154, 196
Meyrpvitz, J.   134
Miami   148, 150
Michigan   26, 106, 141
Microsoft   130, 131, 195
Middle East   73, 171, 179
Midwest   143
migration   23, 47, 78
Milan   146
Miles, I.   16
milieu   67, 187
  of innovation   67–8
mini-cities   124
Minitel   127
Minneapolis/St.Paul   76
minute, calls per subscriber   174
  telephone calls   84
mirror, images   153
  worlds   136
MIT   59
Mitchell, W.M.   137
Mitchelson, R.L.   23, 63, 85, 103–5, 189
MNCs.   79, 102, 210
mobile   195
  communications   197, 203
  information services   195
  telephones   19, 193, 194, 197–8
  telephone technology   196
  telephony   18, 65, 80, 86 152, 188, 195, 198
  see also cellular
mobility   195, 216
modems   140
  calls   83
Montreal   118
Moore, D.   154
MOOs   33, 137
Morgan, K.   163
Morley, D.   162
Morris, M.   24
Morse, S.F.B.   157
Mosco, V.   20, 23, 42
Moss, M.L.   21, 85, 93, 97–8, 103–5, 108,
    144–5, 192
motion pictures   21, 82, 86, 94, 106, 107, 103,
    108, 135, 145, 198, 215
  see also film; movies
movies   4, 9, 10, 19, 41, 107
  see also film; motion pictures
MP3 81, 200
Mapping a Future for Digital Connections   181
MS-Office   136
Mulgan, G.   119
multimedia   4, 54, 94, 104, 108, 114, 131
  games   107
  gulch   98, 105
  messaging (MMS)   198
  Super Corridor, Malaysia   74
  transmission   135
multination region   74

multinational corporations (MNCs)   47, 59, 104,
    120, 122, 126
Munich   146
Murray, A.T.   199
museum   90
music   4, 116, 117, 129
  CDs   81

Namibia   170
NAPs   151
NASDAQ   65, 80, 78, 105, 206
  crisis   71, 79
Nashville   190
nation   75, 123, 174, 183
  states   55, 120
national   21, 33, 43, 65, 95, 96, 109, 125, 140,
    146, 148, 172, 173, 194
  boundaries   15
  capital flows   123
  center   79
  city   193
  dimension   44
  economic system   55
  economy   23, 120, 121, 122
  geographical location   185
  high-tech output   75
  integrated economy   156
  investment   67
  leadership   103, 162
  levels   56, 162, 163, 203
  networks   152
  or regional   66
  origins   65
  policy   65, 119, 175
  regulating bodies   120
  scale   65, 121
  societies   50
  systems of production   74
  telecommunications services   180
  territory   15, 103, 214
  traditions   44
  wealth   173
nationally integrated economy   174
nations   65, 74, 99, 125, 161, 165, 166, 167,
    170, 171, 180, 199
nature   127
Negroponte, N.   34
Neotrace   195
Nepstar   200
NetGeo   195
Netherlands   118, 165, 167, 175
NetNames   89
Netscape   88
NetValue   199
networks   15, 21, 30, 31, 47, 86, 89, 90, 95, 124,
    128, 129, 132, 134, 136, 137, 138, 140,
    141, 143, 152, 153, 155, 183, 186, 187,
    190, 207, 208, 214
  access points   148
  cities   214
  communications   136
  infrastructure   129, 181
  layer   139

of interaction 68
of interconnections 101
of networks 19, 88
relations 61
solution 89, 98
network-based 198
networked 76, 216
cities 42
communications 135
individualism 188
networking 186
culture 190
Networld 134
new economy 57, 80, 206
index 75
New Jersey 207
New Mexico 74
new societies 175
New York 42, 46, 47, 66–7, 75–6, 80, 84, 85,
89, 93, 96, 97, 99, 101, 103–106, 109,
111, 112, 114, 118, 120, 122, 123, 125,
131, 132, 133, 143, 144, 145, 147, 148,
149, 150, 152, 154, 158, 189, 190, 191,
201, 211
New York State 106
New York Stock Exchange 47, 103
New York University 89
New Zealand 165, 175
news 4, 14, 193
agencies 18
newspaper 24, 86, 107, 154, 161, 163, 165,
167, 170–3, 192
Nielsen Media Research 109, 188–9, 191, 193,
201
Nijman, J. 103
Noam, E.M. 19
node 42, 89, 143, 186
Nokia 77, 173
Nolan, R.L. 88
Nordic countries 77, 173, 174, 175, 179, 203,
213
high-tech industry 77
leadership 173
nations 214
societies 76
traditions 76
see also Scandinavia
norms 61, 74
North America 11, 48, 55, 73, 73, 73, 88, 103,
118, 162, 164, 165, 167, 171, 172, 175,
179
North Dakota 150
northern countries 74
Norway 77, 170, 173, 214
NSF 89
NSFNET 88, 141, 143, 148
NTIA 140
NUA 128–9, 132, 163, 169, 172, 177, 179–80
Nye, J.S., Jr. 12, 156, 159

O'Kelly, M.E. 140–1, 144, 148, 152
O'Toole, L. 118
Oakland 152

OECD 3, 114, 158, 198, 200
countries 179, 180, 200
offices 19, 36
offshore 122, 131
Ogan, C. 24
Ogden, M.R. 49
Ohmae, K. 55
Ó hUallacháin 61, 73–6
O'Neill, H.O. 85
online 193, 200
newspaper 193
openness 74, 174–5
operating system 88
operation 17, 27
operational location 96
operators 17, 19, 20, 21, 22, 26, 197
oracle 113, 131
organization 34, 43, 58, 60, 62, 64, 74, 81, 90,
91, 95, 133, 140, 186, 187, 188, 194, 208,
211, 214
application 76
culture 110
framework 32, 33
interaction images 42
structure 19
synergy 67
Osaka 84
Ostbye, H. 173
Oulu 77
outsourcing 60
overnight transfers 85
Owens, W.A. 12, 156
ownership rights 25

Pacific 67, 73, 74, 179, 186, 195, 213;
centers 73
coast 189, 203
rim 122, 162, 165, 166, 167, 170, 172, 173,
175, 176
packaging 22, 91, 93, 103, 109, 110, 153, 212
packet switching 87, 157
Palestinian Authority 179
Palfini, J. 187
Palmer, J.W. 127
Paltridge, S. 116, 176, 179–80
paper 8, 14, 25, 120, 174, 208
Paris 99, 114, 123, 146
region 85
Park, T.M. 199
Parker, E.B. 181
Parr, R.L. 25
Pastore, M. 201
Patel, P. 60
patents 61, 64, 75, 75, 76, 158
attorneys 60
citations 61
laws 25
protection 54
pattern 47, 48, 74, 100, 155, 185, 192, 200,
201, 202
Pavitt, K. 60
Pavlik, J.V. 104

PCs   10, 14, 18, 82, 106, 131, 139, 140, 156–8,
      166–7 170–1, 194, 196–8, 215
Pelletiere, D.   177, 191–2
penetration   154, 161, 162, 162, 163, 166, 167,
      170, 171, 174, 177, 180, 188, 189, 190,
      191, 192, 199, 201, 202, 203, 213
perception   2, 33, 38, 56
peripheral areas   55, 131, 202
peripherality   21
personal communications system   186
personal construction   44
personal voice communications   170
Pew report   187
phantasmagoria   44, 47–8, 183
Philadelphia   76, 103, 191
Philippines   47
Phronesis   6
Pierce, S.   195
place   7, 8, 9, 21, 22, 24, 25, 26, 29, 31, 32, 33,
      34, 36, 37, 39, 43, 44, 46, 47, 48, 49, 50,
      51, 56, 58, 61, 92, 95, 96, 103, 122, 124,
      125, 131, 132, 133, 135, 136, 137, 181,
      182, 194, 195, 198, 209, 212, 213, 214,
      215
   making   136
   production   49
placeless   34, 53
places,
   of investment   78
   of management   120
   of website production   95
   real   30
   sticky   61
planning   198
plants   79
points-of-presence   151
policies   74
politics   1, 26, 105, 173, 183, 206, 208
   and cultural transparency   25
   boundaries   9
   economy   42
   events   123
   space   43
POPs   152
population   93, 103, 144, 162, 163, 171, 179,
      191, 200, 201, 214;
   centers   89
   densities   173
   ranking   97, 99
   size   92, 100, 109, 145, 212
Population Reference Bureau   179
Porat, M.   2, 13, 16
pornography   129
portable information machine   195
portals   117
Portland   189
post   84
postal address   91
post-industrial economy   16
Postman, N.   3
postmodernism   126
power   10, 11, 15, 22, 24, 25, 27, 33, 120, 121,
      159, 172, 186, 208, 209

battles   15
holders   15
of the mind   60
relations   46
struggle   50
Pred, A.   40, 48, 68
primate city   99–100
Principality of Sealand   131
printed   192
documents   82
forms   31
information   14
printers   4, 14, 21, 25, 104
packets,
   priority   181
private ownership   156
privatization   12, 23
process   32, 47, 48, 64, 91, 108, 109, 111, 120,
      122, 126, 194, 203, 206, 211, 213
processing knowledge-expertise   91
processing   10, 21, 22, 27, 91, 92, 105, 111,
      185, 211
producer   5, 40, 51, 92, 141, 161, 209, 211, 213
producer services   24, 54, 60, 67, 77, 104, 123
producers of information   51
product   14, 16, 19, 31, 60, 63, 64, 90, 119, 209
   cycle theory   111
   introduction   75
   life-cycles   531
   marketing   194
   of information   30
   technology   57
production   xi, 8, 10, 12, 15–18, 20–4, 27, 30,
      33, 34, 36, 37, 38, 41, 48, 53, 55, 56,
      61–5, 68, 74, 75, 76, 77, 78, 81, 84, 108,
      113, 114, 118, 120, 125, 132, 136, 141,
      155, 157, 158, 159, 161–3, 179, 182, 188,
      189, 191, 199, 202, 206, 208–14
   decision   63
   factors   12, 65, 67
   firms   79
   force   32
   management   67
   of a website   24
   of electronic consumer goods   158
   of capital   125
   of information technology   213
   of information   2, 23, 24, 113, 124, 182, 213,
      215, 216
   of knowledge   216
   of place   120
   of space   33, 37, 122
   of technology   191
   price   77, 173
   process   3, 64
   relations   110
   scheme   65
productive   79
   capital   120
   organization   64
productivity   10

products 6, 33, 60, 64, 65, 68, 80, 117, 128,
    129, 130, 131, 132, 153, 156, 162, 175,
    183, 194, 202, 208, 214
professional human resources 65
proximity 123, 149, 205, 212
public to the private sector 12
public 12, 156
    networks 15
    relations 76
publishing 25, 105
    company 91
    industry 104
Purcell, D. 120
purchasing power 194
Pusan 154

Qatar 179
quaternary occupations 16
Quay, R. 177
Quova 195

R&D 13, 61, 62, 63, 64, 78, 79, 79, 94, 106, 111,
    114, 124, 156, 195, 206, 209, 210, 210
    180
    activities 57
    basic 58
    frontier 88
    funds 60
    location 68
    processes 62
    region 64
    workers 23
radio 2, 4, 17, 18, 24, 83, 84, 86, 103, 104, 157,
    161, 163-7, 170-2, 174, 182, 192, 195,
    197
railways 55, 86, 141, 171
Raleigh–Durban–Chapel Hill 72
real estate 16, 36, 126, 206
Reed, H. 125
Regan, P.M. 26
region 24, 54, 59, 60, 64, 67, 68, 76, 77, 79,
    118, 122, 123, 125, 161, 162, 171, 172,
    173, 177, 179, 180, 182, 207, 210
    states 55
regional 21, 33, 34, 65, 95, 111, 123, 125, 128,
    140, 162, 175, 203, 213
    blocs 55
    concentration 64
    cooperation 175, 176
    economy 64
    enterprises 181
    fountainhead 77
    industrial organization 61
    ISPs 140
    learning 103
    level 162, 163
    planning 190
    scales 181
    setting 60, 62
    task 61
registration 91
regulation 22, 23, 77, 162, 194
Relph, E. 34, 40

representation 34
    of place 135
    of space 33, 37–38
research 16, 24, 57, 62, 72, 73, 128, 193, 198
    centers 62, 87, 89, 141
    facilities 71
    funds 62
    institutions 65
    knowledge 62
    organization 58
    universities 65
Research Triangle Park 106
resources 2, 5, 6, 16, 32, 33, 33, 55, 67, 119,
    209
restaurants 198
retail trade 128, 158, 194
re-territorialization 48
Rheingold, H. 137
Richardson, R. 181
Richtel, M. 207
RIPE 186
Rivlin 116
Roberts, J. 2, 4, 7, 9, 26
Roberts, S. 122
Robertson, R. 44
Robins, K. 16, 29, 34, 44, 162
Rodrigo, G.C. 177, 191-2
Romero, S. 197
Roos, J.P. 173-4
Rosencrance, R. 214
Roszak, T. 1
Russia 151

Sack, R. 32
sales 129
    processes 6
    tax 194
Salvaggio, J.L. 13
San Diego 181, 189, 201
San Fernando Valley 118
San Francisco 31, 54, 66, 72, 75, 89, 96, 97, 98,
    99, 105, 108, 109, 110, 118, 131, 132,
    137, 143, 144, 145, 147, 148, 149, 150,
    151, 152, 159, 189, 190, 193, 201, 205,
    206, 206, 211, 212, 214
San Jose 54, 72, 75, 98, 108, 110, 133, 145,
    151, 189, 206, 211, 214
Santa Monica 108
Sassen, S. 27, 41, 92, 101-2, 104, 120-3, 125
satellites 18, 31, 139, 197-8
Saudi Arabia 138, 170
Saxenian, A. 74, 76, 110, 211-2
Saxony (Germany) 74
Scandinavia 76, 172
Scandinavian 167, 171, 172
    countries 77, 173
    leadership 173
    nation 74, 165, 167, 170, 214
    see also Nordic
scanners 14
Scarborough 193
Schement, J.R. 10, 12
Schiller, H.I. 12

Schivelbusch, W. 55–6
Schrag, Z.M. 19, 136–8
science park 59
scientific 54
    knowledge 62
    material 116
    technical workers 71
    technological information 67
scientists 78
South China Morning Post 208
Scott, A.J. 64, 94, 102, 108–9
search 82
    engines 114, 116–7, 128
Seattle 76, 89, 131, 133, 143, 144, 189, 201
security 77, 79, 80, 153, 207
    related R&D 98
sense, of place 40, 49, 135
Seoul 96, 97, 100, 154, 191, 199, 202
separation 215
server farms 86, 131, 132, 138, 140, 149, 152,
    205, 206, 215
servers 153
services 6, 16, 21, 23, 26, 36, 54, 57–8, 78, 107,
    108, 117, 122–4, 128, 129, 130, 131, 140,
    141, 162, 163, 166, 181, 183, 192–3,
    194–8, 205, 207, 214
    areas 21
    economy 12, 16, 122, 155
    providers 14, 15, 77, 132, 174
sex 114, 116–8, 190
Shamai, S. 33
Shanghai 100, 192, 195
Shefer, A. 68
Sheppard, E. 30
Shields, R. 49
Shiode, N. 186
shipping 125, 175
Shiva, V. 43
shopping 86–7, 114, 117, 192
Short, J.R. 41, 95, 104
Siemens 163, 164
signal flows 40
Silicon Alley 105
Silicon Valley 65, 67, 72, 74, 75, 76, 106, 110,
    126, 152, 153, 194, 206, 212
Silverstone, R. 163
Simonsen, K. 31
Simpson, J.A. 39
simulation 135, 138
simultaneity 50
Sinclair, T.J. 43
Singapore 24, 65, 122, 126, 131, 138, 165–7,
    170, 172, 175, 196, 199
sites 34, 50, 88, 94–5, 114, 117, 119, 132, 136,
    143, 153
skilled workers 71
Slough 132
smart buildings 36
Smith, D.F., Jr. 66
Smith, G.V. 25
Smith, M.P. 43, 50
Smith, N. 55

SMS 198, 203, 213
social 47, 54, 56, 105, 172, 176, 180, 182, 185,
    188, 190, 202, 208, 209
    action 3
    activity 14, 215
    actors 15
    and cartographic space 30
    and cultural areas 14
    and cultural aspects 9
    and cultural dimensions xi
    and economic 27, 40
        institutions 68
    appreciation 60
    attitudes 56
    bonds 137
    capital 61
    communities 137
    connectivity 123
    contacts 14, 123, 134, 137, 175
    context 134, 203
    culture 40, 54
    digital divide 181
    formation 135
    geography of information 23
    hierarchy 15
    indicators 214
    influences 46
    interaction 200
    level 187
    life 44, 46, 135, 206
    lives 36
    networks 137, 186, 187
    or economic processes 5
    organization 10, 50
    patterns 47
    practices 42
    process 42
    realities 137
    relation 31, 33, 40, 42, 67
    resource 31
    scales 33
    space 31, 36–8, 46, 122, 137
    strata 76
    structure 22, 40
    technologization 54
    theory 33
    uses 186
    value 32, 34
social-cultural characteristics 175
socialization 6
societal 185
    elements 11
    transmissions 10
society 2, 10, 13, 22, 33, 44, 54, 54, 55, 76,
    139, 155, 156, 158, 177, 182, 187, 188,
    208, 213, 215
    programmed 12
socioeconomic 182, 185, 189, 192, 206
    access 22
    constraint 35
    levels 105
    systems 12
sociospatial 180

classification 33
context 187
software 10, 14, 15, 20, 53, 64, 88, 103, 105,
        106, 109, 111, 129, 157, 158, 211
Sohonet 99
Soja, E.W. 32, 106
Sölvell 62
SOMA 207
Sophia Antipolis 74
South Africa 73, 170
South America 47, 48, 122
South Dakota 150
Southern California 106
South Korea 165
sovereignty 214
Soviet Union 78
space 7, 8, 14, 15, 22, 29, 30, 31, 39, 40, 41,
        43, 44, 46, 47, 49, 50, 51, 55, 56, 119,
        122, 124, 124, 125, 126, 133, 135, 136,
        137, 138, 149, 174, 205, 209, 213, 215
    of flows 15, 42, 56–7, 161, 181–2, 214–215
    of imagination 37
    of places 15, 43, 56, 161, 181, 182, 213–4
    of representation 38, 33
    reorganization of 56
space-based 56
space-consuming 196
spaceless places 49
Spain 99
spatial 26, 30, 125
    accent 31
    agglomeration 42
    anchoring 137
    and temporal significance 55
    and temporal worlds 46l
    barriers 56, 57, 125, 126
    change 55
    communities 187
    dimension 9, 46
    discontinuity 68
    effects 33
    expressions 33
    extent 187
    fixity 36
    interactions 187
    language 135, 137
    metaphors 137
    national barriers 120
    networking 21
    organization 54
    patterns 21, 47
    perspective 7, 181
    practices 33, 38
    scale 33, 50
    splintering 182
    structuring 138
    study xi
spatially 23, 40, 216
specialization 13, 68, 76, 93, 94, 97, 101, 103,
        107, 108, 110, 111, 113, 114, 117, 125,
        162, 182, 211, 213, 214, 216
speed 19, 34, 40, 67, 128, 141, 181, 198
spillover, of tacit knowledge 78

Sprint 141
Sproul, L.S. 187
star scientist 58
start-up 212
    activities 79
    companies 23, 78
    enterprises 65
    high-tech 108
state 19, 22, 23, 74, 118, 128, 150, 174, 182,
        194, 214
    policies 176
    restrictions 41
statism 14
Steinfeld, C. 13
Sternberg, R. 61, 66
Stewart, T.A. 57
Stockholm 72, 77
stock market 44, 122, 206
Storper, M. 3, 7, 59, 106–8
Stough, R.R. 61
subscribers 156, 197, 201
suburbia 80
Sudan 179
Swartz, J. 181
Sweden 65, 163–7, 170, 172, 173, 214
SWIFT 121, 124
switching 158, 207
    facilities 148
    points 212
    system 151
    tools 88
Switzerland 88, 164, 165, 167, 172, 175
Swyngedouw, E.A. 32, 37, 43–4, 121, 124–6
symbol 7, 32, 33, 37, 42, 134, 135, 137
    city 137
    landscapes 136
    manipulation 15
    places 136, 137
    system 7
    territoriality 137
Syria 179
system 87, 88, 89, 116, 196, 213
    integration 19
    of information ownership 87

tacit knowledge 7, 9, 23, 59, 61, 74, 77, 79, 109,
        209, 211, 216
    spillover 62
Taipei 73
Taiwan 154, 199
Tanner, K. 23
taxation 26, 60, 194
Taylor, S. 133
TCP/IP 85, 88, 139
TDL 90
Taejon 154
techne 6
technological 19, 174
    activities 60
    advances 15
    communications media 213
    conditions 10
    development 20, 119, 156, 174

technological (*continued*)
  incubators 79
  infrastructure 22, 68
  innovation xi, 58, 60, 63, 75, 155
  lead 156
  network 41
  potential 163
  products 64, 207
  revolution 26
  sophistication 190
  standards 174
  tacit knowledge 54
technology xi, 7, 8, 9, 18, 19, 20–22, 29, 30, 34,
      42, 53, 57, 67, 68, 76, 88, 105, 107, 109,
      111, 114, 118, 125, 136, 141, 153, 157,
      162, 172, 176, 177, 185, 187, 189,
      194–196, 200–202, 206, 207, 212
  and flows 54
  assessment 64
  business 76
  center 109
  crisis 153
  development 146
  effort 197
  measures 75
  new 64, 210
  production centers 75
  production 189
  R&D 214
  socialization 54
technopoles 64–5, 67, 131
Technopolis 65, 67
technoscapes 41
tech-poles 72, 75, 188
Tel Aviv, metropolitan 79
Telecom Finland 173
telecom hotels 149
telecommunications xi, 10, 13, 14, 19, 21, 23,
      29, 34, 36, 46, 55, 64, 67, 81, 87, 93, 94,
      103, 104, 109, 111, 114, 119, 120, 123,
      124, 127, 132, 141, 148, 155, 156–59,
      162, 174, 175, 177, 179, 206, 208, 210,
      211, 214, 215
  business 77
  channels 5, 6
  company 21, 173
  equipment 149
  hubs 149
  industry 65, 76, 156, 175
  infrastructure 54, 95, 104, 119, 138, 152,
      162, 183, 212, 214
  junction 170
  lines 140
  means 156
  media 55
  networks 6, 16, 139
  revolution 156
  services 156
  system 99, 139, 156, 162, 207
  technologies 57, 85, 155, 186
Telecommunications Act 21
TeleGeography 140–1, 146–7, 149, 154–5
telegraph 55, 56, 155, 157

telehouse 139, 149, 207
telematics 93
telephone 4, 8, 10, 14, 15, 17, 18, 19, 20, 77,
      84, 86, 87, 139, 141, 155, 157, 158, 161,
      163, 164, 165–7, 170–3, 179, 182, 185,
      187, 191, 195, 198, 207, 213
  calls 5, 18, 46, 85, 137, 139, 154, 156
  companies 20, 188, 200
  copper wires 200
  exchanges 139
  hub 175
  infrastructures 171
  lines 139, 141, 195
  modems 198
  numbers 197
  operating companies 21
  picture 157, 202
  sales 129
  service 20
  switchboards 157
  system 20, 87, 127, 139, 170
  technology 171
  traffic 83, 104, 154
telephony 86, 117, 154, 156, 173
teleports 155
television 2, 4, 10, 15, 17–8, 41, 83, 84, 86, 91,
      94, 103, 107, 108, 109, 127, 134, 135,
      154, 157, 161, 163–7, 170–2, 174, 182,
      185, 191, 192, 195, 204, 207, 213
  industry 145, 156
  programs 14, 22, 24, 83
  sets 14
  stations 18, 24
telex 157
territory 26, 61, 74, 135, 137, 174, 175
terror 116
test laboratories 60
Texas 74, 89, 143, 144
text 5, 6, 14, 21, 32, 33, 58, 84, 86
textile products 47
Thailand 208
theater 107
third generation (3G) 196
Thomson Financial Investor Relations 104
Thrift, N. 3, 40–1, 44, 46, 49, 55, 57, 120, 123
Thu Nguyen, D. 24
time 7, 8, 12, 14, 15, 25, 33, 35, 46, 55–7, 122,
      135, 138, 170, 181, 185, 192
  -based society 56
  budget 192
  length 195
  perspective 74
  saving 198;
  sharing 42
  use 197
time–space, compression 44, 46, 49, 57
  constraints 188
  convergence 46, 56
  distanciation 44
Time-Warner 20
Tivers, J. 134

Tokyo Internet   140
Tokyo   46, 84, 85, 93, 97, 99, 101, 102, 104, 120, 122, 123, 125, 147, 148, 154
top-level domains (TLDs)   89
Top US Book Markets   106
Tropical Guinea   179
Toronto   97
Touraine, A.   12
tourism   114, 125, 126, 134
Townsend, A.   21, 88–9, 93, 97–8, 105, 108, 142–5, 150, 154, 158, 171, 192, 194–5, 214
trade   16, 59, 102, 123, 175, 194
traffic   83, 113, 152–4, 206–7
  congestion   153
transactions   57, 86, 162, 187, 187, 193, 194, 207
transatlantic, backbone capacity   154
transfer   37, 46, 86, 105, 111, 148, 210, 216
  of capital   125
  of information   128
transformation, to digital economy   75
transistor technology   5, 164
transit   140, 48, 153
transition   158, 170
transmission   xi, 8, 10, 14, 15, 17, 19, 21, 22, 27, 37, 47, 53, 55, 79, 84, 87, 102, 110, 111, 139, 140, 158, 159, 181, 188, 197, 198, 205, 207, 208, 209, 210, 211;
  and consumption   53
  and use   14
  capacity   95
  channels   18
  device   5
  information   155
  lines   139
  media   5, 18, 155
  nodes   155
  of capital   6
  of information   8, 9, 13, 55, 88, 113
  of knowledge   9, 25, 59
  of phone calls   20
  of telephone   20
  system   8, 18, 19, 54, 89, 212
  towers   152
transnational, economic investments   155
  urbanism   50
transportation   49, 55, 103, 165
  hubs   132
  innovations   56
  networks   141
  technologies   56
travel   114, 117, 129, 136
  agencies   37
Trondheim   77
trust   61, 74, 77, 123
Tseng, K.F.   21
Tuan, Y.F.   136
Tuathail, G.O.   43, 44
Tunisia   73
TV Basics [website]   107
Tyner, J.A.   47

UK   46, 85, 90, 98, 99, 118, 121, 132, 133, 157, 158, 159, 165, 181, 196
  see also British
US   11–6, 19, 20, 24, 47, 57, 61, 63, 67, 74, 75, 77, 78, 84, 85, 88–90, 97, 98, 100, 101, 106–9, 118, 121, 127, 129, 130, 132, 133, 141, 146, 147–9, 150–2, 154–9, 162–7, 170, 172–5, 179, 181, 185, 188–93, 196–201, 203, 206, 208, 213–4
  army   87–8
  census   187
  -centric Internet   155
  cities   97, 99, 148
  courts   87
  economy   158
  geographical sources   66
  IT regions   72
  military   87, 90
  national center   104
  outgoing   85
  population   85
  SMAs   93
  telephone calls   83
  transcontinental   88
Uimonen, T.   192, 196
Ullman, E.   32
United Arab Emirates   170
United Nations   121, 163, 164, 166, 177
University, of California   8, 81, 88
  of Haifa   91
  of Illinois   88
  of Michigan   141
  of Texas   127, 128, 129, 131–2
university   8, 13, 23, 58, 61, 62, 64, 71, 72, 73, 74, 76, 87, 88, 89, 128, 152, 215
Unwin, T.   43
urban   21, 33, 102, 124, 124, 163, 181
  and economic geographies   30
  area   34, 77, 108, 152, 174, 189, 201
  centers   95
  concentration   97, 119, 162
  distribution   98, 118, 152;
  economies   125
  elites   179, 180
  entities   175
  financial centers   123
  geography   97, 188, 198, 216
  hierarchy   111, 145, 159, 216
  information   198
  landmarks   136
  landscape   120
  life   205
  markets   188, 201
  modernity   207
  office landscape   126
  pattern   191, 202
  perspective   97
  population   146, 201
  regions   89
  size   216
  social theory   50
  space   33, 181
  spatial fixity   162

urban (*continued*)
  specialization 21
  structure 146, 199
  systems 119, 181
urbanism 92
urban–rural 181
URL 88
Urry, J. 54, 134
use 15, 21, 25, 33–36, 58, 64, 100, 127, 133,
    135, 138, 154, 155, 156, 161, 163, 176,
    177, 180, 183, 185, 187, 190, 192, 193,
    196, 197, 200, 202, 208, 209, 213
Usenet 83, 187
user 18, 20, 22, 30, 34, 35, 49, 91, 92, 94, 95,
    118, 119, 136, 140, 153, 164, 165, 186,
    187, 192, 193, 195–7

Valentine, G. 25, 36
values 134
Varian, H.R. 81–4, 136, 192
vendors 132
venture capital 54, 57, 60, 62–8, 71–3, 79, 80,
    108, 110, 206, 207, 210, 212
ventures, new 71
video 109, 118, 199, 200;
  adult 214
  conferencing 18, 200, 202
  games 4, 192
  information 198
  phone 18, 20, 202
Virginia 106
virtual 29, 30, 31, 33, 34, 50, 91, 118, 135, 188,
    206, 209
  spaces 38
  casinos 26
  communities 31, 136, 137
  -electronic 128
  geography 30, 134
  image 27, 137
  interaction 34
  meeting place 41
  places 40, 120
  presentations 135
  space production 135
  space 30, 31, 33, 34, 36, 37, 38, 49, 56
  spatial images 209
  states 214
virtuality 135
viruses 154, 208
visual, electronic entertainment 107
  electronic information 186
  route 195
visualized interaction 140
voice 4, 5, 14, 21, 86, 198
  communications 195, 200
  telephone calls 154
  telephony 6
VoIP 139

war 116, 120
Warf, B. 21–2, 26, 88, 120, 176, 179–81, 192
Washington Post 193

Washington, DC 76, 96, 98, 143–5, 148, 149,
    189, 190, 214
Watts, M.J. 40, 48
W-CDMA technology 196
Web 5, 18, 24, 86–8, 91, 96, 97 105, 112–5,
    117, 128, 129, 133, 135, 136, 138, 177,
    189, 190, 198, 208, 212
  address 89
  communication 19
  deep 83
  hosting companies 140
  hosting facilities 152
  hotels 152, 153
  information 87, 110, 115, 210, 215
  pages 141
  production 190
  surface 83
Web-based environments 124
Web-enabled databases 131
Web site 22, 26, 33, 47, 83, 86–7, 92, 94, 95,
    96, 98, 103, 110–1, 113–4, 131–3, 136,
    140, 152–3, 208, 210, 215
  hosting 153–4
    facility 153
    servers 152
  hosts 140, 150
  information 5, 153
  locations 195
  -owning firms 98
  producers 153
  production 118
Webster, F. 10
Weibull, L. 173
Weiner, E.S.C. 39
welfare 172
WELL 31, 137, 190
Wellman, B. 137, 179, 181, 187
Wells, P. 158
West-Coast cities 201
West Germany 100
Western cities 190
Western Europe 121, 151, 172, 175
Westland, J.C. 127, 133
Wheeler, J.O. 85, 103–5, 141, 145, 189
wholesale 128
WiFi (wireless fidelity) 196
Wigand, R.T. 127–8
Williams, H. 34, 162
Wilson, M.I. 26, 89, 91, 118–9, 179–80
wired 68, 188
  telephony 195
wireless 158
  communications 205
  information systems 86
  information transmission 193, 195
  laptops 194
  mobile telephones 186
  networks 196, 197
  services 196
  telephony 195, 207
  transmission 198
  waves 139
wires 85

WLAN   196
Wolff, G.   41
work   15, 42, 172
   culture   67
   force   131
   place   195
   practices   3
   style   76
workers   65, 71, 106
World Bank   121, 168–9, 171
world capacity   153
world city   41, 103, 131, 146
WorldCom   141
world culture   103
WorldPaper   214
World Trade Center (WTC)   80, 207
World War II   158

World Wide Web (WWW)   31, 49, 83, 128, 161, 171
Wurster, T.S.   16
Wyoming   150

XDSL   198–202

Yahoo!   116, 133, 194
Yale University   88
Yellow Pages   198
Yeoh, B.S.A.   43
Yeung, H.W.   131, 175

Zaire   179
Zook, M.A.   5, 63–4, 67, 81, 89–92, 94–8, 105,
      108, 111, 114, 117–8, 189
Zurich   123, 146